Blockchain in the Industrial Internet of Things

IOP Series in Next Generation Computing

Series editors

About the series

The motive of this series is to develop a trusted library on advanced computational methods, technologies and their applications.

This series focuses on the latest developments in next generation computing, and in particular on the synergy between computer science and other disciplines. Books in the series will explore new developments in various disciplines that are relevant for computational perspective including foundations, systems, innovative applications and other research contributions related to the overall design of computational tools, models and algorithms that are relevant for the respective domain. It encompasses research and development in artificial intelligence, machine learning, block chain technology, quantum cryptography, quantum computing, nanoscience, bioscience-based sensors, IoT applications, nature inspired algorithms, computer vision, bioinformatics etc. and their applications in the areas of science, engineering, business and social sciences. It covers a broad spectrum of applications in the community, from industry, government, and academia.

The aim of the series is to provide an opportunity for prospective researchers and experts to publish works based on next generation computing and its diverse applications. It also provides a data-sharing platform that will bring together international researchers, professionals and academics. This series brings together thought leaders, researchers, industry practitioners, and potential users of different disciplines to develop new trends and opportunities, exchange ideas and practices related to advanced computational methods and promote interdisciplinary knowledge.

Blockchain in the Industrial Internet of Things

Lakshmana Kumar Ramasamy
Hindusthan College of Engineering and Technology, Coimbatore, India

Seifedine Kadry
Faculty of Applied Computing and Technology (FACT), Noroff University College, Kristiansand, Norway

IOP Publishing, Bristol, UK

ISBN 978-0-7503-3663-5 (ebook)
ISBN 978-0-7503-3661-1 (print)
ISBN 978-0-7503-3664-2 (myPrint)
ISBN 978-0-7503-3662-8 (mobi)

DOI 10.1088/978-0-7503-3663-5

Version: 20210501

IOP ebooks

British Library Cataloguing-in-Publication Data: A catalogue record for this book is available from the British Library.

Published by IOP Publishing, wholly owned by The Institute of Physics, London

IOP Publishing, Temple Circus, Temple Way, Bristol, BS1 6HG, UK

US Office: IOP Publishing, Inc., 190 North Independence Mall West, Suite 601, Philadelphia, PA 19106, USA

Contents

Preface

Blockchain is a continuously-growing ledger that maintains a constant record of all business dealings in a safe, period sequence and unchanging manner. It could utilize for the safe exchange of cash, assets, agreements, and so on. The Industrial Internet of Things (IIoT) is the expansion and utilization of the Internet of Things (IoT) in industrial applications and areas. Effectually, it is the integration of Industry 4.0, which seems to pay more attention to the performance of the industrial progression. In this book, we did our best to cover comprehensively these two topics and the benefits of their combination. In chapter 1, we define IoT with the discussion of different areas of its applications. Chapter 2, discusses the overview, working process, advantages, applications, challenges, and futures of IIoT. The debate on blockchain's overview, its structure, and how it is executed in different aspects is provided in chapter 3. In chapter 4, we show the six essential blockchain key characteristics. Chapter 5 debates the smart contract's overview, its structure, how it is executed and its different aspects. The taxonomy of blockchain for Artificial Intelligence (AI) classified concerning decentralization of AI algorithms and functions, blockchain type and formation, and the basic consensus algorithms used for decentralized dispersed transaction verifications over vital networks, is given in chapter 6. Chapter 7 investigates the combination of blockchain with IIoT. In chapter 8, we examine the latest model for linking IIoT to the blockchain. We name the combination of blockchain and IIoT as the Blockchain of Industrial Internet of Things (BIIoT). The structure of BIIoT proposed in this chapter as well. The deployment discussion of BIIoT was given in chapter 9. The last chapter is dedicated to BIIoT industrial applications.

The authors

Acknowledgements

Acknowledgements of Lakshmana Kumar Ramasamy

First I would like to thank the Almighty for helping me in writing a book. This book would not have been possible without the cooperation of Hindusthan College of Engineering and Technology that allowed me to develop and test insight-related ideas in projects, workshops, and consulting engagements over the last eight-plus years. I want to acknowledge the management of Hindusthan Institutions, who have constantly encouraged me to 'get this book done.' I cannot thank them enough for being my unrelenting source of inspiration to challenge how things get done in the academic world. I am eternally grateful to my parents and father in law, who taught me discipline, tough love, manners, respect, and so much more that has helped me succeed in life. Finally, I want to thank my wife, Jayanthi, a lifelong partner who makes both the journey and destination worthwhile and especially my kids, Nirantharaa and Mhaghizhan for tolerating my incessant disappearances into my home office. An additional thanks to the IOP family; I'm deeply indebted for their wonderful editorial support and guidance.

Acknowledgements of Seifedine Kadry

First and foremost, I would like to praise and thank God, the Almighty, who has granted countless blessings and knowledge. I want to thank the IOP family widely for this opportunity; I am particularly grateful to the Series Editor Professor Prateek Agrawal, and to John Navas and Sarah Armstrong for all the support they gave us, without them this project would still be in process. I want to thank my family, my lovely wife Diana and my little angel Hadi. Special thanks go to my teachers and supervisors to whom I owe a lot.

Author biographies

Lakshmana Kumar Ramasamy

Dr Lakshmana Kumar Ramasamy is currently associated with Hindusthan College of Engineering and Technology, India. He is also a Director-Research & Development (Artificial Intelligence) in a Canadian based company (ASIQC) in Vancouver region of British Columbia, Canada. He is the Founding Member of IEEE SIG of Big Data for Cyber Security and Privacy, IEEE. He is a global chapter lead for MLCS [Machine Learning for Cyber Security]. He himself is involved in research and expertise in AI and Blockchain technologies. He holds the Data Science certification from John Hopkins University, United States. He also holds the Amazon Cloud Architect certification from Amazon Web Services. He holds the privileged Gold level partnership award from Infosys for bridging the gap between industry and academia in 2017. He holds the privileged Silver level partnership award from Infosys for bridging the gap between industry and academia in 2016. He is an Editor-In-Chief for the Book Series, 'Advances in Quantum Computing, Artificial Intelligence, and Data Sciences for Industrial Transformation Series', and also Editor-In-Chief for Focus Series, 'Convergence of IoT and Blockchain Through Machine Learning Approaches', CRC Press—Taylor & Francis Group, LLC, USA. He is a core member in Artificial Intelligence Editorial Advisory Board of Cambridge Scholars Publishing, UK. He is an Editorial board member of *Trends in Renewable Energy Journal*, USA and *Frontiers in Communications and Networks*, Switzerland. He is the Associate Editor in *International Journal of Quality Control and Standards in Science and Engineering* (IJQCSSE), IGI Publishers, USA. As part of his professional career, he has around 50+ publications in international journals and conferences. He has participated in eight edited books from Elsevier, CRC Press, Springer and Wiley. He holds around 19 Indian patents and two international patents. He is an ACM distinguished speaker and IEEE brand ambassador.

Seifedine Kadry

Dr Seifedine Kadry has a Bachelor degree in 1999 from Lebanese University, MS degree in 2002 from Reims University (France) and EPFL (Lausanne), PhD in 2007 from Blaise Pascal University (France), HDR degree (Habilitation) in 2017 from Rouen University. At present, his research focuses on data science, education using technology, system prognostics, stochastic systems, and applied mathematics. He is an ABET program evaluator for computing, and ABET program evaluator for Engineering Tech. He is an IET Fellow. Currently, he is a full professor of data science at Noroff University College, Norway.

IOP Publishing

Lakshmana Kumar Ramasamy and Seifedine Kadry

Chapter 1

Internet of things (IoT)

The Internet of Things (IoT) refers to the evolutionary stage of the internet, which makes a global communicating infrastructure between humans and machines. IoT is constructing the global infrastructure which will change the fundamental aspects of our lives, from health services to manufacturing, from agriculture to mining. IoT will offer the necessary facilities of the latest rising artificial intelligence (AI) development. This chapter discusses the overview, characteristics, advantages, disadvantages, common uses, security, trust, privacy and functional view of IoT. Furthermore, this chapter proposes application areas of IoT in detail.

1.1 IoT overview

IoT has grown to be a marketing trend and general news piece. Beyond exaggeration, IoT appeared as a powerful technique with appliances in numerous domains. IoT has origins in multiple former methods: sensor networks, embedded systems and pervasive informatics. Many IoT devices are linked mutually to develop specific purpose schemes; in the global network, they are rarely utilized as public access devices.

An IoT node is a sensor contained hardware piece that broadcasts sensed information to users or any other devices over the internet. IoT nodes embed into industrial equipment, mobile and medical instruments, wireless sensors, and more. Top examples of IoTs are connected smart city, smart industry, smart transport, smart buildings [1], smart energy, smart manufacturing, smart environment monitoring, smart living, smart health, smart food and water monitoring. Figure 1.1 shows the IoT network architecture. This architecture has a lot of IoT sensors for sensing purposes such as temperature, humidity, pressure etc. After sensing, these data are transmitted to a cloud server via an IoT gateway. Furthermore, users can access these data through mobile apps and so on.

Due to the accessibility of low cost and smart devices, the IoT network refers to a smart system. IoT devices operate independently with their hearing and transmission

doi:10.1088/978-0-7503-3663-5ch1

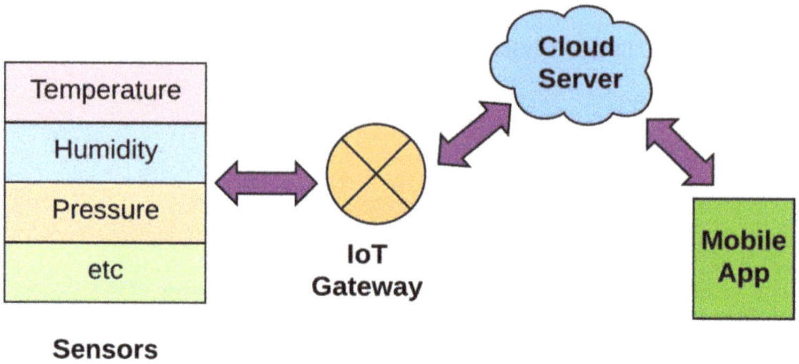

Figure 1.1. IoT network architecture.

abilities. Furthermore, the propagation of IoT provides a lot of benefits but also provides potential threats. An overlooked factor so far is the rise in energy expenditure. IoT nodes are anticipated to always be accessible on other nodes. IoT offers a lot of benefits, including:

- **Locating and tracing abilities**: Customers should be capable of tracking the nodes and locating them in a short amount of time.
- **Ubiquitous information swap**: In IoT where nodes are linked to the internet and where information is transmitted. Ubiquitous means intelligence. Therefore, intelligence sensors collect information and transmit it using a prearranged input.
- **Enhanced power solution**: Customers should be capable of tracking even the strongest node, and the customer should be capable of obtaining the best result.
- **Data and intelligence management**: IoT does not always require providing commands to the instrument; where the node gives intelligence and information previously it can start working and obtains decisions and discovers solutions based on intelligence.
- **Scalability**: IoT should be the measurability, as with any number of IoT nodes above an extensive network all nodes should distinguish uniquely.

Also, numerous significant IoT problems can be identified. These open problems make it clear that the complexity of internet design currently needs significant capabilities to alter.

- **Unprotected authorization/authentication**: The administrator usually presents authentication to verify the customer identity, and the authorization utilizes rewriting or modifying the content for that appliance and the consent that the administrator will give.
- **The technology of server**: The number of IoT nodes over the IoT field increases the demand and the number of IoT node replies, moreover increases simultaneously depending entirely on the server where customers use the

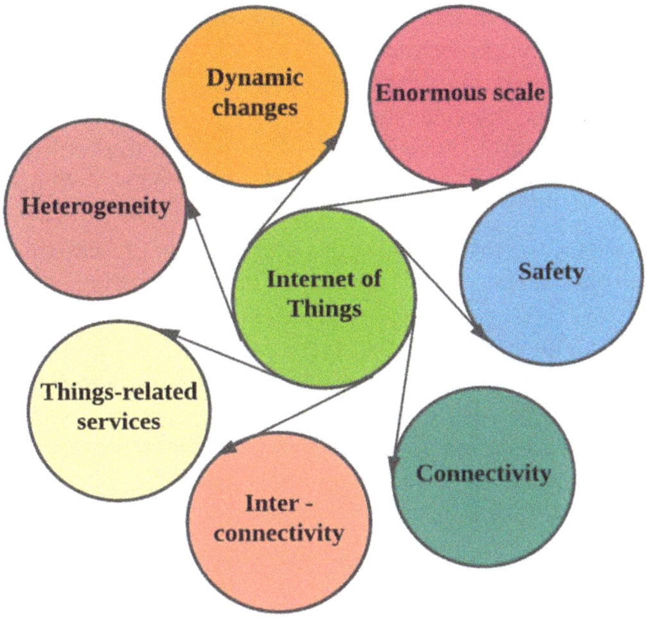

Figurere 1.2. Characteristics of IoT.

interface. The server response to the IoT node demand should be made immediately. There must be no delay in responding to the customer.

- **Management of storage**: A massive quantity of information is created. When connected IoT nodes have a massive quantity of multimedia data transmitted, they have big data and other types of inconsistent files where data is held concerning these IoT nodes, these files do not take much space. Still, many of them should be useable as soon as possible.
- **Data management**: As transmission between nodes is completed, more information is created daily between nodes, and there is more information to be transmitted from one location to another. Consideration should be given to whether specific information is transmitted or not.
- **Security**: Provision of security can be challenging as the automation of nodes has increased, which has generated novel security problems.

1.1.1 IoT essential characteristics

Figure 1.2 shows IoT essential characteristics discussed as follows:

- **Enormous scale**: The number of IoT nodes that require handling and connecting will be a order of magnitude higher than the IoT nodes currently linked to the internet. Of particular importance will be the administration of the information created and its analysis for appliance reasons. It is about content semantics, with content management.
- **Safety**: As customers achieve advantages from IoT, customers should not fail to remember security. As senders and receivers of IoT, they should plan for

security. This encompasses the security of customer information and the security of customer welfare: protecting networks, endpoints, and is a universal message that represents generating a measurable safety model.

- **Dynamic changes**: Device status changes drastically, for instance, sleep and wakefulness, connection and disconnection and content of IoT nodes contain speed and position. Besides, the number of IoT nodes can modify energetically.
- **Heterogeneity**: Various IoT nodes depending on various networks and hardware platforms. They should connect to other nodes via various networks.
- **Connectivity**: This permits IoT to have ease of access and compatibility. Ease of access is obtainable on the network though compatibility, presenting the same capabilities for using and producing content.
- **Services related to things**: IoT can present a lot of services related to things within things restrictions. To present services related to things within things limitations, both techniques in the global and data world will alter.
- **Interconnectivity**: In terms of IoT, anything could be linked to universal data and contact the IoT basic organizational and physical structures.

1.1.2 IoT benefits

Currently, every part of business and lifestyle hopes to realise the benefits of IoT. Figure 1.3 shows a list of a few of the benefits that IoT will provide:

- **Technical enhancement**: Similar techniques and data that enhance consumer observation of IoT facts and enhance IoT node usage, and facilitate the most significant advances in technique. IoT opens up a world of actual data performance and field performance.
- **Enhanced consumer engagement**: Recent statistics have the problem of ambiguity and fundamental errors in precision; also, as mentioned,

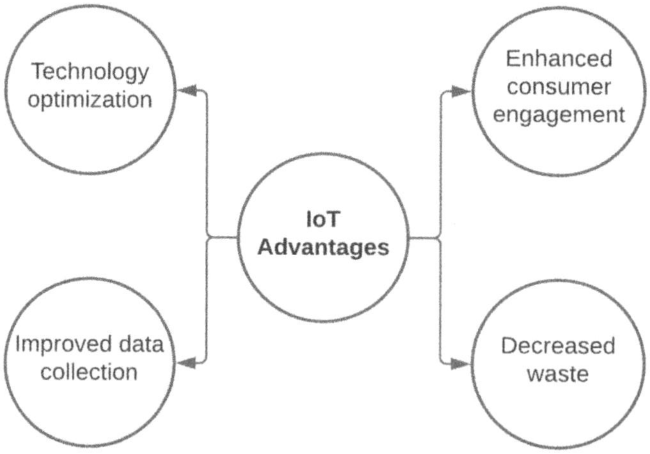

Figure 1.3. Advantages of IoT.

engagement remains are inactive. IoT changes this, attaining a rich and productive engagement, including the spectator.

- **Advanced information compilation**: Today's information compilation undergoes restrictions in plans for practical usage. IoT smashes it down into those gaps and then puts it right where people desire for investigating our planet.
- **Decreased waste**: IoT generates development fields more clearly. Recent statistics provide us with insignificant intelligence, rather IoT presents actuality data that leads to efficient resource management.

1.1.3 IoT drawbacks

IoT provides an inspiring collection of advantages; it moreover provides an essential collection of challenges. Figure 1.4 shows a list of some of its main problems:

- **Security**: IoT generates an environmental scheme for frequently linked nodes contacting networks. Furthermore, the scheme provides minimal regulation despite safety measurements. It protects consumers from different types of hackers [1].
- **Privacy**: IoT expertise presents private information with complete information, not including the involvement of the consumer.
- **Flexibility**: Consumers are worried regarding the elasticity of the IoT scheme for easy integration. The anxiety is about discovering themselves with too many contradictory or protected source codes.
- **Compliance**: IoT, similar to all other technologies in the trade sector, should obey the rules. Its difficulty creates the problem of compatibility seeming like a daunting challenge when many think that typical software compatibility is a war.

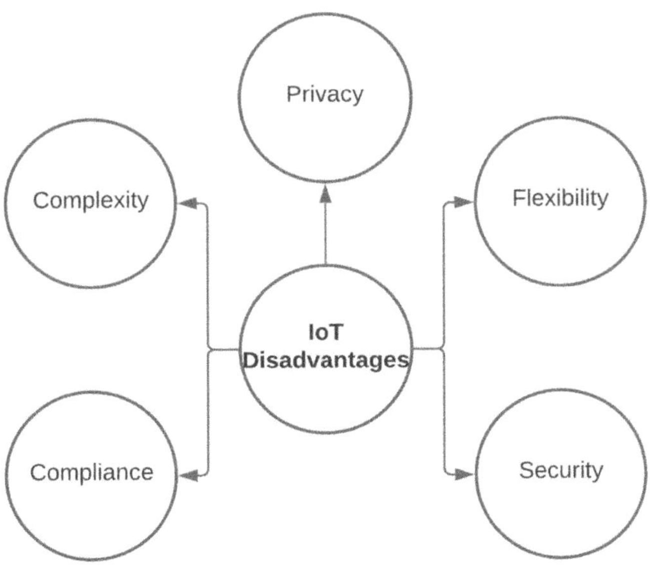

Figure 1.4. Disadvantages of IoT.

- **Complexity**: A few consumers discover that IoT schemes are complicated in the rule of schema, ordering, and storage provided for their usage of numerous techniques and a massive collection of newly permitted techniques.

1.2 IoT common uses

IoT schemes are helpful for many types of appliances:

- Industrial schemes utilize sensors to monitor together with the industry procedures themselves—product excellence—and the condition of the apparatus. A growing number of electric motors, for instance, contain sensors that gather information utilized to forecast future motor breakdowns.
- Smart buildings utilize sensors to discover the positions of persons and the condition of a building. That information can be utilized to regulate ventilation/air conditioning and lighting schemes to decrease working prices. Smart buildings also utilize sensors to monitor the physical condition of the building.
- Smart cities utilize sensors to monitor persons walking rather than travelling in a vehicle as well as vehicle traffic, and can compile information from smart buildings.
- Vehicles utilize network sensors to monitor vehicle condition and offer enhancement, decrease energy expenditure, and reduce inferior discharges.
- Medical schemes link with a variety of patient monitoring sensors that can be situated at home, in an emergency vehicle, or in a hospital.

There are lots of use cases that assist users in recognizing the needs of the IoT scheme.

Notification system: Messages from IoT devices can be collected and examined. Notices made when certain conditions are met.

Sensor network: The scheme can work definitely as an information collection scheme for the sensor sets.

Reactive system: Study of IoT device sensed information could incite actuators to be accelerated. Users retain a word reactive for schemes that do not execute standard regulatory rules.

Analysis system: Messages from IoT devices are collected and examined, other than in that event, the research is continuing. Research outcomes can be created occasionally.

Event latency: Delays from capturing an event to its receiver cannot be significant for volume-based apps but are significant for online research.

Control scheme: IoT device sensed information is nourishing to regulate the instructions that produce the effects for the actuator. Users may discover the category of non-functional necessities that execute to most IoT schemes. Non-functional necessities in the scheme force non-functional needs on the elements.

Buffer volume and event loss rate: If there are no strict limits on event manufacturing standards, the surroundings can generate multiple events over some time over the scheme. The event loss rate holds the preferred abilities, while buffer volume is a practical need that may directly link to the strength of the elements.

Availability and reliability: As IoT, schemes are dispersed; availability is often utilized to explain dispersed schemes. Reliability may determine across network components instead of total scheme reliability.

Throughput and service latency: Finally, procedures would be executed through maintenance. Users could indicate throughput and delays of maintenance.

A lifetime of service: IoT schemes are always anticipated to contain longevity greater than users anticipate for PC schemes. The lifespan of a scheme or subdivision of a scheme might be longer than that of an element, mainly if a scheme utilizes passive sensors and different elements.

1.3 IoT—security

Security is ultimately seen as a significant necessity for each kind of computer scheme, containing IoT schemes. However, most IoT schemes are significantly safer than standard Windows/Linux/Mac schemes. IoT safety issues arise from a variety of reasons: insufficient hardware safety elements, inadequately created software with broad limitations of susceptibilities, and different safety creation faults.

Unprotected IoT devices generate safety issues for the whole IoT scheme. Since devices generally contain a lifespan of numerous years, an enormous established foundation of vulnerable IoT nodes would cause safety issues in the future. Unprotected IoT schemes create safety issues across the internet. IoT nodes abound; unprotected IoT devices are perfectly suitable for a lot of attacks (especially denial-of-service (DoS)). Consumer details are secure from direct theft; however, the IoT needs to be built, thereby low confidentiality information could not be merely utilized for conjecture about high confidentiality information.

Example: A German article states that attackers have hacked the safety scheme. They disturbed the management scheme, which prevented the furnace from closing correctly, resulting in serious harm. Therefore, users may know the consequence of the assault earlier, determining a suitable defence.

Challenges: Aside from pricing and availability of IoT nodes everywhere, different safety problems cause continual trouble to the IoT shown in figure 1.5:

- **Device similarity**: IoT nodes are well homogeneous. These nodes use a similar communication technique and elements. If one scheme or node is vulnerable from susceptibility, numerous others contain a similar problem.
- **Unexpected activities**: The vast amount of IoT nodes used and their vast list of empowering techniques denotes that these nodes activities in the area may be unexpected. A particular scheme might be well-designed within management systems; however, there are no assurances of how it will communicate with other nodes.
- **Device longevity**: One of the advantages of IoT nodes is long life, but that long life denotes that they can live longer with their node assistance. One can measure the similarity or dissimilarity between this and conventional schemes that have to assist and modernize after a long time, with numerous terminating

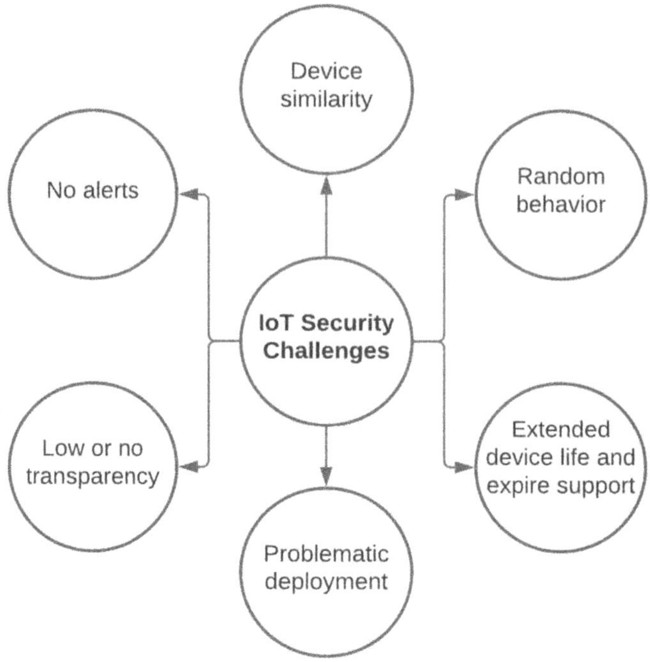

Figure 1.5. Security challenges of IoT.

their utilization. Abandonware and Orphan nodes do not have a similar safety toughness for different schemes because of the emergence of techniques over time.

- **Complex deployment**: One of the primary objectives of IoT is to put superior networks and research where they could not go before. Unexpectedly, this makes a difficulty for physical protection of nodes in these areas that are extreme or only accessible with difficulty.
- **Lack of transparency**: Numerous IoT nodes fail to present clarity in terms of their performance. Consumers are unable to see or use their procedures and can only assume the nodes work correctly. They cannot regulate unnecessary activities or information gathering; moreover, when the producer updates the node, it might carry unnecessary activities.
- **No warnings**: One more purpose of IoT is to offer the best performance in the absence of interruption. This initiates consumer consciousness issues. Consumers are not aware of the nodes or recognize when an unfamiliar thing becomes invalid. Safety violations may continue for a long time in the absence of discovery.

1.3.1 Trust for IoT

A trustworthy structure needs to be able to manage people and equipment as consumers, e.g. it needs to transfer hope to people and to be strongly sufficient to be utilized by types of equipment without DoS attack [2]. The building of hope structures that tackle this necessity will need improvements in fields, for example:

- Lightweight public key infrastructures (PKIs) as a basis for confidence management. Improvements are predicted in hierarchy and cross-certification ideas to allow decisions to deal with the scalable necessities.
- Lightweight key management schemes to allow confidence relations to be set up and to share encryption tools using minimal contact and procedure tools, as well as limited resource nature of numerous IoT nodes.
- Data quality is a necessity for numerous IoT-designed schemes where a set of data that describes data about other information can be utilized to offer IoT information reliable estimation.
- Decentralized and self-configuring schemes are substituted by PKI to set up reliability, for instance, a federated identity, Peer-to-peer (P2P) network.
- Reliability negotiation is a process that permits two parties to discuss routinely, basically a sequence of reliability tenets, the low condition of reliability needed to allow an act of assistance or a data strip.
- Guaranteed techniques for reliable sites contain protocols, software, hardware, etc.
- Access control to avoid information violations. One instance is usage regulation, which is the procedure of guaranteeing the proper use of specific data according to a pre-defined principle, after which the access to data is allowed [3].

1.3.2 IoT privacy

Most of the data in an IoT scheme can be private information; there is a need to assist in anonymous and limited private data management. There are many fields where development is needed:

- Cryptographic technologies that allow secured data to be saved and progressed and distributed, without the data being accessible to third parties.
- Securing the secrecy of the area, where the area can be incorporated into things related to human beings [3, 4].
- Restriction of private data presumption, people may desire to maintain it confidential, by looking at IoT associated transactions.
- Maintaining data as the neighborhood as is feasible utilizing key management and distributed computing.
- A soft identifier application, where consumer identity can be utilized to make a wide variety of different identities for particular appliances. One soft identifier may not build for a particular theme without exposing unnecessary details, which could lead to a violation of confidentiality [4].

1.4 Functional view of IoT

The IoT system indicates to uniquely discoverable devices with their visual presentations in a structure such as the Internet and IoT problem solving that include many elements, for example:

1) IoT local device communication model (a small distance wireless network can be contacted, for instance, installed on a mobile phone or situated in the surroundings of a consumer etc). This model is in responsible for acquiring, monitoring and transmission to remote servers for perpetual storage and research.
2) Local research module and observation processing obtained by IoT nodes.
3) Model to communicate with remote IoT nodes,directly on the internet or possibly by proxy. This model is in charge of observation acquired and transmission to remote servers for perpetual storage and research.
4) Data analytics applications model works on an application server that serves all customers. It takes demands from the web and mobile customers as well as appropriate IoT views as input, implements relevant information executing algorithms and provides output based on skills that will be provided after to consumers.
5) Consumer interface (mobile or web): measurements of the visual representation in the given circumstances and consumer communication, e.g. description of customer enquiries.

1.5 Application domains

IoT application areas identified by the IoT European Research Cluster (IERC) based on the inputs of specialists, reports and studies. Top examples of IoT applications are smart city, smart industry, smart transport, smart buildings, smart energy, smart manufacturing, smart environment monitoring, smart living, smart health, smart food and water monitoring.

Industrial automation and production stressed from the shortlived manufacturing life cycle and the need for short-term marketing in numerous domains. The next-generation of production schemes will be constructed on elasticity and redesigned as an important goal. The new list of IoT appliances presented beneath includes examples of IoT applications in different areas, indicating why IoT is one of the planned technique fashions over the subsequent five years.

1.5.1 Smart industry

Machine-to-machine appliances: Automatic machine diagnostics and asset management.

The liquid level of a tank or vessel: Monitoring fuel and water levels in storage vessels and wells.

Silo volume calculator: Measure of emptiness and weight of merchandise.

Air quality within and around buildings and structures: Supervising of dangerous gas and oxygen quantities within an industrial process plant that manufactures chemicals to guarantee the safety of workers and merchandise [5, 6].

The presence of ozone: Ozone monitoring for the period of dry meat processing in the food industry.

Measurement and monitoring of temperature: Temperature control inside medical refrigerator systems with important ingredients.

1.5.2 Smart mobility and transport

Shipment Quality: Supervising of container openings, vibrations, impacts, or any damage for insurance.

NFC Payment: Enable merchants to get credit or debit card payments online by offering a link to a merchant bank or acquirer based in a place or time required to complete the activities of public transportation, museums, galleries, and so on.

Object Location: Search individual objects in large locations such as repositories or ports.

Monitor the activity of fleet vehicles and assets: Route control of critical assets such as jewels, medical drugs or hazardous materials.

Non-compatibility discovery of storage: Alerting of containers' emission of readily combustible materials close to others holding explosives.

Car management: Car sharing organizations run the usage of vehicles using smartphones with net connections fixed in each car.

Automatic vehicle diagnosis: Data gathering from CAN Bus to transmit a real-time alarm to immediate risks or to offer tips to a person who drives a vehicle.

1.5.3 Smart buildings

Liquid availability: Liquid discovery in data centres, buildings for storing goods and significant construction underground to avert collapse and the process of slow destruction by chemical action.

Access control of perimeter: Selective restriction of access to protected areas and the discovery of human beings in unauthorized areas [3].

Control of indoor climate: Measuring and controlling a physical quantity that expresses hot and cold, equipment for producing light, carbon dioxide in fresh air at ppm and so on.

Preserving culture through art: Condition monitoring within museums and art galleries.

Intrusion prevention system: Discovery of doors and windows openings and breaches to detect intruders, especially into a building with criminal intent.

Residential-irrigation: Smart irrigation and monitoring system.

1.5.4 Smart energy

Installation of photovoltaic: Supervising and enhancing of operation in solar power stations [7].

Grid modernization: Power usage supervising and controlling.

Wind energy converter: Supervising and examining the flow of power from a wind energy converter, as well as two-way contact with smart meters for customers to examine usage patterns.

Radiation rate: Shared calculation of radiation rates in surrounding nuclear power plants to create leak warnings.

Stream: Measuring water force that pushes water through pipes in water flow schemes.

1.5.5 Smart production

Composting: Temperature and humidity management in hay, alfalfa, straw, and so on to deter fungi and various microbiological contaminations.

Intelligent management of manufactures: Controlling stock rotation based on FIFO (First-IN, First-OUT) rules on warehouse shelving and storage to stock replenishment automation.

Descendant care: Controlled breeding in farm animals to guarantee their health and survival [8].

Measurements of toxic gas: Research of air pollution in farm buildings and the discovery of dangerous gases from stables [5, 6].

Tracking animals: Locating and identifying animals that graze in open grazing lands or area in the largest stables.

Telecommuting: Providing workers with technology to allow local offices will decrease costs, increase productive capacity, and increase job opportunities while decreasing staff housing, reducing office maintenance and cleaning, and removing everyday office travel.

Production-line monitoring: Production-line monitoring and managment based on radio-frequency identification (RFID), sensors, video surveillance, remote data sharing and cloud-based solutions that allow production-line information to change to business-based schemes.

1.5.6 Smart environment monitoring

Air contamination: Decreasing CO2 emissions from industries, vehicles and hazardous gases created in agriculture fields [5, 9].

Discovery of forest fire: Flue gases monitoring and monitoring of preventive fire conditions for the identification of warning areas [10].

Protection against avalanches and landslides: Monitoring of the water stored in the soil, density of the Earth and vibrations to discover harmful patterns in earth states.

Wildlife protection: Tracking calls use GSM/GPS modules to discover and trace wildlife and supply their co-ordinates through SMS.

Premature earthquake discovery: Distributed control in certain quake areas.

Monitoring oceans and coasts: Using a variety of sensors combined in aircraft, ships, satellites etc for maritime security, fishing vessel tracing and unsafe oil supplies etc.

Weather station networks: Research of meteorological conditions in agricultural land to predict the formation of ice, drought and air change.

1.5.7 Smart living

Water and energy consumption: Monitoring energy and water use to get advice on how to save costs and resources.

Smart shopping system: Get advice on where to sell based on consumer preferences, buying behavior, the existence of allergies, or expiration dates.

Remote control devices: Remotely turn on and off devices to prevent accidents and for energy saving.

Smart home products: Transparent LCD display refrigerator that shows what is inside, expired food details, ingredients for a well-stocked kitchen and all the data obtainable through the app on the smartphone [11]. Washing machines permit one to control the laundry from far away and work routinely when electricity prices are very low. Smart cooking apps monitor the automatic cleaning function of the oven and permitting adjustable temperature control from a distance.

Meteorological station: Shows conditions of the outside weather, for example, humidity, heat, atmospheric pressure, air velocity and rainfall using meters capable of transmitting information over long distances.

Regular checking of safety procedures and standards: Baby alarm, an optical tool used to record images, and house alarm schemes that are making human beings feeling safer in their everyday lives at home.

Gas detector: Real-time data concerning gas consumption and the state of gas piping can present by linking local gas meters to the internet protocol (IP) network. Concerning monitoring and assessing water quality, the result could be a diminution in labour and repair prices, enhanced precision and meter readings price reduction, and perhaps gas usage diminution.

1.5.8 Smart health

Activity monitors as support for older persons' physical activity: Wireless body area network measures movement, critical signals, blurred vision and the cellular unit, gathering, displaying and storing activity information.

Falling-detection: Help for the elderly or disabled who live independently.

Pharmacy refrigerators: The conditions of the regulation within cold storage for storing medicines, vaccines and organic components.

Patient monitoring: Surveillance of patient health status within hospitals and nursing homes.

Personal care of sports people: Vital sign monitoring in high-performance areas and camps. Fitness and health manufacture for these reasons are available, that calculate fitness, steps, weight, blood pressure, and other details.

Managing chronic illness: Patient surveillance programs with full patient details can be obtainable for monitoring chronic illness patients remotely. Admission to a reduced medical facility, low price, and temporary hospital stay can be just a few of the advantages.

Hand hygiene approach: Monitoring scheme by using RFID tags for wristbands with the integration of Bluetooth LE tags at the patient's hand hygiene control, where vibration alerts are transmitted to notify during handwashing. The entire information gathered generates statistics that can utilize tracking patient morbidity in specific health staff [8].

Ultraviolet light: Measuring UV sunlight to alert humans not to be exposed at specific times.

Oral care: A Bluetooth-connected dental brush with an app on the smartphone examines the task of brushing teeth and provides details on the smartphone to obtain confidential information or to show statistics to the dental surgeon.

Sleep control: IoT devices situated across the bed sense usual movements, such as breathing and heartbeat and huge movements caused by tossing and turning at night when asleep, presenting information obtainable via the smartphone app.

1.5.9 Smart water monitoring and smart food

Water Leaks: Discovery of water availability in outside tanks and pressure differences in pipelines.

Quality of water: Analysis of the appropriateness of water in natural flowing watercourses and oceans for all of the animal life present in a particular region or time and the appropriateness for potable usage.

Flooding: Monitoring the variability of flow rates of rivers, reservoirs and dams.

Control the supply chain: Monitoring storage conditions during supply chain activities and manufacture tracing for tracking reasons.

Water management: Real-time data concerning water condition and water usage can gather by linking a water meter to the IP network [12].

Enhancement of wine excellence: Monitoring water stored in the soil and trunk width in vineyards to regulate sugar levels in grapes and health of grapevine.

Grounds where the game of golf is played (golf course): Selected irrigation in arid areas to decrease the essential green-water assets.

Greenhouses: Regulate microclimatic level to increase vegetable and fruit manufacture and excellence.

In-field water quality monitoring: Decreasing food spoilage with excellent tracking, statistical management, continuous data acquisition, and crop field management, with excellent management of fertilization, irrigation and electricity.

1.5.10 Smart city

City noise mapping: Monitoring noise in the concentric zone, including bar areas in real-time.

Construction health: Vibration and conditions of materials monitoring in statues, bridges, monuments and popular buildings [8].

Traffic jam: Pedestrian and vehicle level monitoring to enhance walking and driving paths.

Safer cities: Video surveillance system, public address sound systems and fire prevention and control systems.

Intelligent lighting: Smart and weather adaptive lights in raised sources of light on the edge of a road.

IoT applications differ, and IoT applications work for various consumers. Various categories of consumers have various driving requirements. From an IoT point of view, there are three essential consumer categorizations:

- individual citizens;
- groups of citizens (citizens of a nation, state and city);
- companies.

Examples of IoT applications required by individual citizens are such that:

- Increasing their security or the security of members of their family—for instance, remote control for a security alarm system, or activity recognition for older people;
- Facilitating easy-to-perform tasks—for instance: home inventory management reminder.

1.6 Summary

This chapter describes an overview, characteristics, advantages, disadvantages, common uses, security, trust, privacy and functional view of IoT. Furthermore, we propose application areas of IoT in detail. IoT can promote a functional variety of industrial appliances like logistics, manufacturing, food business and services. Novel standards, novel trade, competition, and the need to transport nonstop goods are challenges new businesses face nowadays. As a result, a lot of companies rely on Industrial Internet of Things (IIoT), which refers to any performance executed by businesses to model, supervise and enhance their business processes during insights gathered from thousands of linked machines to assist them in enhancing economical profit. Therefore, the next chapter discusses IIoT in detail.

References

[1] Nia A M and Jha N K 2017 A comprehensive study of the security of Internet-of-Things *IEEE Trans. Emerging Top. Comput.* **5** 586–602

[2] Daud M, Rasiah R and George M 2018 Denial of service: (DoS) impact on sensors *4th Int. Conf. on Info. Management (ICIM)*

[3] Yang Y, Liu X and Deng R H 2018 Lightweight break-glass access control system for healthcare Internet-of-Things *IEEE Trans. on Industrial Informatics* **14** 3610–7

[4] Yang Y, Wu L, Yin G, Li L and Zhao H 2017 A survey on security and privacy issues in internet-of-things *IEEE Internet Things J.* **4** 1250–8

[5] Jamal H, Huzaifa M and Sodunke M A 2019 Smart heat stress and toxic gases monitoring instrument with a developed graphical user interface using IoT *Int. Conf. on Electrical, Commun., and Computer Engineering (ICECCE)*

[6] Kodali R K and Rajanarayanan S C 2019 IoT based indoor air quality monitoring system *Int. Conf. on Wireless Commun. Signal Processing and Networking (WiSPNET)*

[7] Barman B K, Yadav S N and Kumar S 2018 IOT based smart energy meter for efficient energy utilization in smart grid *2nd Int. Conf. on Power, Energy and Environment: Towards Smart Technology (ICEPE)*

[8] Wu F, Wu T and Yuce M R 2019 Design and implementation of a wearable sensor network system for IoT-connected safety and health application *IEEE 5th World Forum on the Internet of Things (WF-IoT)*

[9] Muthukumar S, Sherine Mary W and Jayanthi S 2018 IoT based air pollution monitoring and control system *Int. Conf. on Inventive Res. in Comput. Appl. (ICIRCA)*

[10] Prabha B 2019 An IoT based efficient fire supervision monitoring and alerting system *Third Int. Conf. on I-SMAC (IoT in Social, Mobile, Analytics and Cloud)*

[11] Agarwal K, Agarwal A and Misra G 2019 Review and performance analysis on wireless smart home and home automation using IoT *Third Int. Conf. on I-SMAC*
[12] Rajurkar C, Prabaharan S R S and Muthulakshmi S 2017 IoT based water management *IEEE Int. Conf. on Nextgen Electronic Technologies: Silicon to Software (ICNETS2)*

Chapter 2

Industrial Internet of Things

The Industrial Internet of Things (IIoT) is the expansion and utilization of the Internet of Things (IoT) in industrial applications and areas. Effectively, it is the integration of Industrie 4.0, which seems to pay more attention to the performance of the industrial progression. IoT is a kind of network that links materials or things by a wireless or wired network, in contrast, IIOT is a type of network that handles complicated physical devices connected to sensors used in industrial automation and its associated software. This chapter discusses the overview, working process, advantages, applications, challenges and future of IIoT.

2.1 Industrial IoT overview

IIoT is a primary trend that has important impacts on the World's financial markets. It comprises factories indicating 62% of Gross Domestic Product (GDP) within G20 countries, according to Oxford's Department of Economics, including production, farming, mining and services. It includes industries that rely on tangible physical products to run a business, for example, companies that run hospitals, harbours and storehouses or that provide transport, logistics and health facilities. Of course, the potential benefits of IIoT are massive [1]. The very conservative autonomous assessments put global spending on IIoT at $20 billion in 2012, with spending estimated to attain $500 billion at the end of 2020. Functional capacity is one of the main attractions of IIoT, and primary adopters concentrated on these advantages. By introducing largely automatic equipment in a system of the manufacturing process with more flexibility, for example, producers can increase their production by 30%. Predictable asset maintenance is one of the focus areas, saves up to 12% more than planned repairs, reducing repair costs by up to 30% and removing damage by up to 70%. For instance, Thames Water is the biggest company providing sewage and water services in the UK [2]. It uses sensors, analysis and actual-time information to assist sewer and water service companies predict tool repairs and reply quickly to crisis circumstances, for example, leakages or extreme climate changes. However, there is

doi:10.1088/978-0-7503-3663-5ch2

still a lot to the story. While companies are looking and starting to use IIoT as a performance-enhancing tactic, it provides rich power to those who manufacture tools and production to initiate profitable digital products, creating completely new income streams to enhance bottom and top lines. Based on the operators and owners of the tool, for example, companies in the process of manufacturing, there is an actual chance for raised incomes.

An easy win is to evade the resting time and the closure of facilities and plants, thus increasing productivity. For instance, the petrochemical manufacturer may depend on speculative care to avoid unnecessary closure and maintain manufactures flowing. Apache Corporation, a company engaged in hydrocarbon research, uses this method to forecast offshore and onshore oil pump failure to assist reducing lost manufacture. Officials at Apache say that if the international oil industry could enhance pump process by 1%, it will raise oil manufacture by five hundred thousand barrels per day and obtain for the industry $19 billion per year. Or consider a mining process, where the capability to rapidly carry out a metal test when drill bits unpredictably strike a hard rock allows miners to start the job in less time than previously. However, a few trailblazer companies are moving forward with novel means to utilize IIoT to offer value to consumers, including ZF Friedrichshafen AG, CLAAS KGaA mbH, and Michelin [3]. These companies point to encouraging instances; it is significant to identify that IIoT provides completely new methods of making and thinking about manufacture and services.

2.2 How does the IIoT system work?

IIoT is a group of equipment that has its computing capacity linked to building systems that monitor, gather, interchange and examine information [4]. Each IIoT environment consists of:

- smart devices that can detect contact and retain data about it;
- private and public communication system controlled by numerous telecommunication and broadcasting services
- applications and analysis that create trade data from source data; and
- human beings.

Smart and Edge devices send data directly to the network infrastructure, where it is changed into functional data on how a particular part of the equipment works, for example. This data can be utilized for predictive-maintenance, also, to improve trade procedures.

2.3 IIoT advantages

One of the important advantages an IIoT gives businesses is predictive maintenance [5]. Companies could utilize actual-time information created from IIoT to forecast machine faults, for instance, before they happen, activating organizations to take steps to deal with these problems before a machine part breaks or the machine fails.

One more general advantage is enhanced fieldwork. IIoT technology assists fieldwork professionals in discovering possible problems with consumer machines before they begin to be important problems, allowing the technology to solve issues before they disrupt consumers.

The process of tracking physical assets is one more benefit of IIoT. A party that supplies services or merchandise to producers and consumers could utilize the software for asset management to trace product status and location across the supply chain. The scheme will transmit immediate warnings to stakeholders if the merchandise is spoiled or in danger of being spoiled, providing them with an opportunity to take instant or precautionary action to rectify the circumstance.

Furthermore, IIoT allows improved consumer contentment. When manufacture is linked to the IoT, a producer could obtain and examine information concerning how consumers utilize their manufacture, allowing manufacturers and product developers to design upcoming IoT equipment and construct roadmaps with consumers in mind.

Additionally, IIoT enhances facilities management. Production machinery is vulnerable to damage caused by everyday use, which could be worsened by some situations within the industry. Sensors can monitor light, heat, motion and sound that may lead to less efficient operating situations.

IIoT solutions permit a very capable, inexpensive and simple method to maintain procedures. The utilization of intelligent sensors and appropriate software will permit accurate predictions of future failures. The device could be returned and stored in a very capable way, reducing the situation involving exposure to the danger of mechanical failures and industrial procedures have stopped. The limits of IoT applications are large and varied. Currently, solutions of IoT can be used to do all of the following shown in figure 2.1.

Resource Management: Condition-based maintenance (CBM) is a non-functional activity, so the sensors could be utilized to maximize the effectuality of resource

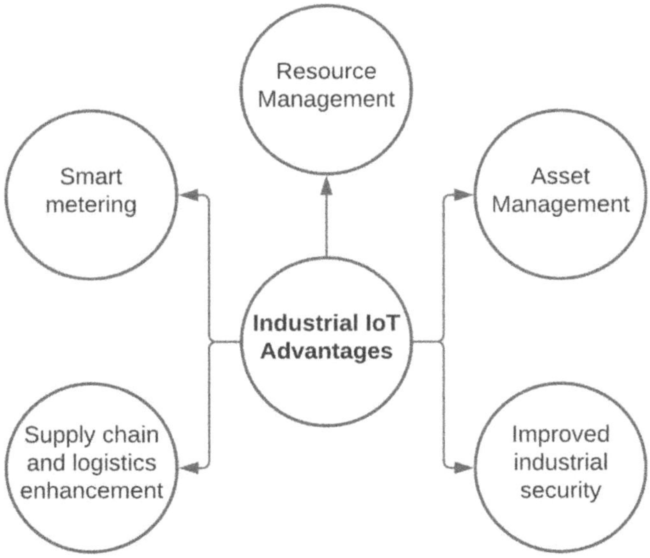

Figure 2.1. Industrial IoT advantages.

management. A production device tends to crack. It is further vulnerable to certain conditions. Sensors could supervise heat, vibration, and other elements that can lead to suboptimal working conditions.

Asset Management: Solutions of IoT could further be very useful when it comes to decreasing exposure to the danger of asset management faults. Events would be effortlessly supervised from one side to the other of the supply chain, providing industries with a complete overview of the asset. Ratings of obtainable goods and supplies can be summarised, which avoids recessions.

Improved industrial security: IoT is well integrated with big data analytics. Key performance indicators of security and health can be regularly supervised to guarantee superior operating conditions. Delayed indicators such as the number of unfortunate incidents could be determined quickly.

Supply chain and logistics enhancement: Actual-time supply chain data would become obtainable by dependence on solutions of IoT. Manufacture and goods would be very easy to trace, and recessions and incompetence could be indicated. In a sense, plants would link to cloud providers. All stakeholders could track the necessary data to forecast problems and deal with the asset diminution as soon as feasible.

Smart metering: Smart meters could supervise the use of assets such as fuel, water, electricity, and so on. By using sensors of IoT, producers would recognize usage and what to do. With effective management, the operating cost could substantially decrease.

2.4 IIoT versus IoT differences

Although IoT and IIoT possess numerous similar technologies, including sensors, cloud platforms, data analysis and machine-to-machine (M2M) communication, they are utilized for different intentions [6].

Applications of IoT link devices in numerous vertical ways, including healthcare, mobility, retail, agriculture, cities and society, education and skills training etc. IoT devices include smart devices, straps and various appliances that usually do not generate emergencies when something goes wrong.

Applications of IIoT, contrastingly, link devices and equipment to industries, for example, energy, utilities and mining industry etc. System downtime and failure in IIoT positions can lead to life-threatening or high-risk circumstances. Applications of IIoT are more concerned with enhancing performance and enhancing safety and health, compared to the user-centricity of IoT appliances.

2.5 IIoT vendors

There are many vendors with IIoT platform [7], including:

- Predix Platform by GE Digital, a company, offers IIoT services and software to industrial companies;
- IoT system by Cisco, an IT and networking company;
- ABB Ability by ASEA Brown Boveri, a power, heavy electrical equipment, and automation technology company;

- Wonderware distribution by Schneider Electric, energy and automation digital solutions for efficiency and sustainability company;
- FIELD system by Fanuc, an automation products and services company;
- Connected Building Software by Honeywell, a technology and manufacturing company.

2.6 Industrial IoT applications

The development of IIoT transforms almost every aspect of today's industry. Figure 2.2 demonstrates the four main assorted areas of IIoT application: maintenance, infrastructure, and supply chain and process control [8].

2.6.1 Maintenance

Maintenance is a very important aspect of the efficient operation of a production line. Efficient preservation decreases idle time and can reduce power utilization, particularly for high energy consumption machines such as engines that could consume excessive power if used in a poor situation. Usually, preservation methods fall into two divisions: preventive and reactive.

Preventive maintenance makes a plan of preservation tasks that are pursued regardless to conditions of machinery, which brings consistent preservation and labour cost with less immediate fixes and longer machinery lifespan; generally, preventive maintenance is assessed to save around 12%–18% of all expenses. Reactive maintenance echoes the feeling, 'if it ain't broke, don't fix it'. Reactive maintenance relies on run-to-failure (RTF) maintenance. In reactive maintenance,

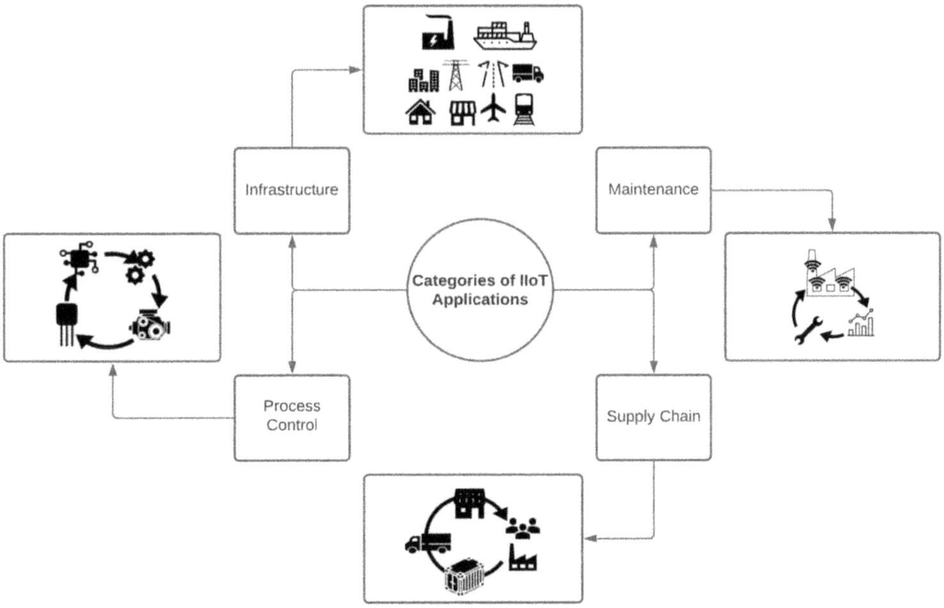

Figure 2.2. Categories of IIoT applications.

maintenance is only executed when machinery has failed. So it decreases maintenance cost-effectively.

IIoT allows predictive maintenance with improved knowledge of machinery conditions due to the distributed sensor technology positioned across all plant floors. Predictive maintenance utilizes sensors to observe and check infrared radiation (IR) and unusual temperature to plan maintenance from forecasting issues from early warning signs. Furthermore, predictive-maintenance saves 8%–12% of all expenses. The drawback of predictive maintenance is that the turnkey cost in technology required to collect forecasting data could be huge. IIoT assists in decreasing a few of those costs by leveraging sensors for commodity and big data analysis to change initial costs.

2.6.2 Infrastructure

Smart devices could create a highly pliable, dependable and capable public organizational structure. These enhancements could bring significant economic benefits for industrial customers of smart organizational structure industries by presenting security improvements, decreasing costs, and reducing people's efforts to maintain and control organizational structure facilities.

For instance, the technology of IoT could integrate information associated with power consumption and, through a smart grid, transmit that data to the internet to analyze usage and cost-cutting advice. One more instance is in the development of smart cities where the technology of IoT could enhance the flow of traffic, make it easier to park, observe and check levels of contamination, and more. A smart city could benefit industries by decreasing travel expenses and making it easier to comply with regulations. One more major market for IoT utilization in the organizational structure sector is in the domain of building automation. Workplaces and houses are nowadays observed and checked with IoT technology. Sensors could observe and check power expenditure, enable settings such as equipment for producing light, a system to maintain a cool atmosphere in warm conditions, and improve safety by monitoring and controlling physical-access. Workplace observing, checking and enhancing could also decrease costs related to ancillary care in assisting with manual labour.

2.6.3 Supply chain

Surveillance sensors, for example radio-frequency identification (RFID), are common in distribution networks for the transportation system, supplier and retailer to observe and check manufactured goods as they cross the supply chain. IIoT provides the ability to greatly enlarge the opportunity of supply chain tracking technologies by offering scope for gathering, distributing, examining, and monitoring goods as they go by various industries, even across international boundaries. By providing real-time data regarding location and transport of merchandise, IoT has the possibility to reduce theft, fraud and other offences. The purpose of a complete supply-chain-management solution (SCMS) is the capacity to monitor goods from the production line in every way by withdrawing from service; although it can monitor goods from manufacturing to point of sale.

2.6.4 Process control

Another aspect of significant development in IIoT is the positioning of sensors in industries to track and adjust a process and monitoring plant to make sure the industry is working properly and to reduce defects quickly, thus growing earnings. The growth of big data analysis to forecast upcoming declines is an important tool for decreasing plant rest periods. Increased automation and monitoring skills decrease the cost and risk of hiring people in manufacturing features. A major challenge to adopting IIoT in plant activities is the requirement for actual time IoT for sensor-to-sensor communication, sensor-to-regulators communication and sensor-to-gateways communication. Advances in concurrent wireless sensor networks and real-time wired (or wireless) sensor networks are of great importance for the route ahead.

2.7 Industrial IoT challenges

This section concentrates on the various challenges industries confront in IIoT and debates moving forward concerning operations enhancement and cost reduction. Figure 2.3 demonstrates the six main IIoT challenges: security, interoperability, scalability, reliability, network performance and management [9].

2.7.1 Security

Challenge: The risk of cyber attacks is higher now than ever before as a lot of industrial users expect to remotely access all things and services based on the cloud

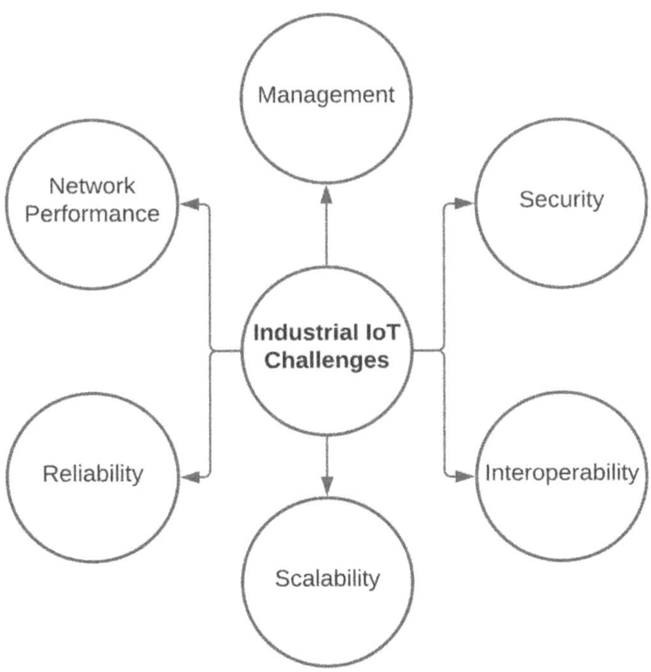

Figure 2.3. Industrial IoT challenges.

linked to their internet. Before now, cybersecurity concentrated on a certain quantity of endpoints. With the arrival of the IIoT, security must increase its attention to comprise virtual and physical worlds on a large-scale.

Solution: The reconstruction of the latest safety structures that take over the whole cyber-physical stack, from device status verification and safety of application to system verification, is not open to discussion. Generally, a solution of IoT is only as strong as its weakest element. For that reason, it is important to strengthen safety with things installed with firewall security that complies with industry internet security levels such as IEC-62443. Since firewalls may differ more in their capability to monitor, store, and report function handled by the device, they should guarantee the device provides login and notification functions. Regarding securing remote access, firewalls must come with VPN support. Internet Protocol Security (IPSec) and Layer Two Tunneling Protocol (LZTP) [10] are also reliable functions for protecting remote access to things. In terms of physical safety, operators can block blind areas by using commercially operated IP tracking cameras on networks installed with Video Always-On technique to maintain the surveillance constantly, whether it works in difficult environments or using one-to-many streaming.

2.7.2 Interoperability

Challenge: In any inter-connected systems, all elements in the equipment can contact all others—speaking a similar language, even if these elements operate in various fields [11]. Lack of standard software connectors and standard connection protocols is a complex thing in the IoT environment. In industry, this indicates that 40% of the total economic value (TEV) of IIoT will stay locked since various systems cannot work jointly. Further, the unrestricted interoperability drive is also difficult for long-term conventional equipment that needs expensive replacement or reuse to operate with the current technology.

Solution: Closing gaps in contact between equipment and the internet will need gateway or middleware to guarantee interoperability across the network by translating information from one protocol to another. For instance, open platform communications (OPC)-enabled contact serves as a gateway that permits information distribution from one element to another of networks without relational limitations, defeating the interoperability challenge. The OPC server could contact information without interruption between programmable logic controller stores and supervisory control and data acquisition programs, remote terminal unit on-premises, human–machine interface (HMI) channels, and software appliances on desktop computers. At the same time, when it comes to legacy to IP based solutions, middleware plays a very important part in making painless IP implementation. The positive news is that the customer does not even need to change the available equipment on their network; therefore, it saves a lot of money for the customer. Besides, it is important that, with the number of network equipment and protocols increasing rapidly, gateway brings faster protocol transformation. It is an area that network developers must regard if they need to combine various protocols effectually.

2.7.3 Scalability

Challenge: With billions of devices previously linked and billions more to come, the big truth is that millions of possible event defects are coming with a huge number of additional devices. Industrial plans must prepare for the construction of an impressive infrastructure ready for expansion. The International Data Corporation (IDC) forecasted that the whole quantity of 'Linked Things' would be more than 200 billion by the end of 2020. And, hereafter, most positioned devices will be mobile, constantly linked, and use less energy—and they will be required to interrelate with their environmental change.

Solution: IoT systems require being flexible and measurable using software or additional performance that easily combines into a complete solution. Latest retrenchment applications of scientific knowledge, for example, the Turbo Chain permit powerful measurability to assist IIoT application; they guarantee that large-scale networks are dependable at all times. On mobile networks, long term evolution (LTE) planned to offer improved abilities for a huge quantity of mobile devices to achieve actual-time connectivity or significant information transactions. Furthermore, low-energy devices fixed with intelligent CPUs could very easily create network development as they contain adequate processing capabilities and work with power proficiency.

2.7.4 Reliability

Challenge: The success of IoT relies on trusted IP networks. Also, the key to a creative and productive network is tried and tested devices. The reliability of the device in the IoT is very important in difficult work environments such as the gas, oil, maritime, energy and rail industries. These factories would depend heavily on remote access, so it is beneficial for them to position sufficient devices. Even if they are well-constructed and easy-to-use, the user interface would not help the customer if the background of the system is unreliable. Deformed devices, or a mechanical failure because of high temperatures and further extreme surrounding factors, would represent a negative impact on IoT adoption. We all know intelligently about the potential consequences of service failure: people at risk, expensive idle time, and incorrect information analytics, giving only these as examples.

Solution: The success of the IIoT relies largely on the powerful infrastructure that works best in the back end. Investing in high-performance industrial devices—including broad working temperature range, high MTBF, four levels of EMS security, dual power supply, and dual LAN technique—applies to industries working in severe circumstances.

Additionally, many downsizing technologies exist to guarantee faster network retrieval. But, a nonproprietary protocol such as rapid spanning tree protocol (RSTP) is not as quick as proprietary-protocols. A trade-off is that open protocols do not apply to each other. For the last couple of years, the advancement of demolition technologies, for example, the Turbo Chain and the Turbo Ring, has allowed collaboration with RSTP and further networks, bringing the most excellent of both worlds.

2.7.5 Network performance

Challenge: Many networks are not designed to meet the challenges of a growing IoT. The practice focuses on the combination of multiple systems and video monitoring; therefore, it requires combined broadband networks to integrate control commands, information, voice and video. Legacy tracking devices have precedent extended to the range. Networks will be under too much pressure to offer increased bandwidth as millions of devices will generate a larger number of data points; all that requires gathering and examining. Moving this quantity of information online will use the latest bandwidth levels; apart from that, networks will suffer from bottlenecks.

Solution: Data-transfer rate is an important deliberation when choosing a network technology for a specified application. Technologies, for example, 5G and 4G (LTE-A, LTE) are excellent for applications of IoT, specified for their big data-transfer rates.

There are many more choices to construct a network with an exponential future. The backbone of the Gigabit standard is a solution for turnaround network engineers to tackle requirements for seamless high-performance. Not to be disregarded is the standard IEEE 802.11n—a growth that provides as far as 300 Mbps and multiple-input multiple-output for uninterrupted video streaming over wireless networks. Besides, the technology of multicast should be available in key components, edge components, and network access layer to enhance the functions, particularly in the most important multi-service networks. Performance of a network could also be enhanced by possessing video programs to make good use of the network with smart video analytics that discovers and begins video streaming by the movement and discovery of objects or warning areas.

2.7.6 Management

Challenge: IIoT requires constructing with preservation and rehabilitation. When creating a complex web of devices, connection system operators handle not only the original system but also handle every latest application.

Engineers emcounter a shallow learning curve. Many access management software human–computer interactions are not planned for labour-saving technology; various management jobs need various devices. To aggravate the situation, using personnel to streamline the network can lead to long working hours and human error. Besides, additional costs are engaged in hiring/training engineers in control centres to reply to faults in a timely manner.

Solution: Network performance management software for modernization makes network engineers more efficient by providing the following tools:

- a combined network-management package that includes installing, operating, maintaining and diagnosing;
- production tool to repair and distribute devices quickly;
- predictive maintenance that includes the use of sensors to supervise equipment often to evoid deterioration and to decide when preservation will be needed, instead of depending on often planned preservation procedures;
- apps based mobile device management that permits developers to examine the actual-time condition of devices or to receive event announcements;
- application programming interface (API) is a strong interface for computing to facilitate the growth of an application or program management.

2.8 IIoT futures

The growth of IIoT will shortly take the industry from expectations to the real world [12].

Asset management, production, manufacture sharing, and transaction management would each be automatic to an extent in the following years. However, to move the digital transition to the forefront, managers must engage in this procedure from the outset. The IT sector would be required to assist in creating revolutionary results around manufacturing procedures.

Bain and company forecasted IIoT appliances would grow to be worth higher than $300 billion by 2020, becoming twice that of the IoT customer category ($150 billion). Likewise, IDC announced the topmost three companies investing in IIoT in 2018 are manufacturing ($189 billion), with attention on wealth management and transport ($85 billion), with attention on cargo monitoring, shipping and services ($73 billion), with attention on smart grids, while customer IoT consumption will attain $62 billion. With high hopes, Accenture plc anticipates IIoT to inject $14.2 trillion to the economic system simultaneously, increasing by 7.3% Compound Annual Growth Rate, known as CAGR by 2020.

The IIoT, named Industrie 4.0, has previously begun the 4th Industrial Revolution (4IR). Higher automation has previously led to a 30% rise in productivity and the obtainability of flexible manufacturing techniques. Maintainance is also forecasted to reduce costs by preventing more than 12% of planned faults and reducing breakdowns by almost 70%.

Experts recommend that in the future, IIoT will also improve manufacturing levels and lead to a variety of innovations. The labour component will be further converted as a segment of the comprehensive automation procedure. The lineup of adoption challenges still requires to be overcome, with forecast analytics predicting that by 2020 there would be 50 million linked devices across the globe. It will be unfortunate if such a large network could not always be used to improve industrial procedures. Keep in mind that IIoT is not regarding intelligent product growth.

Instead, it would result in a high stage of effectiveness and forecasting instead of responsive intervention—the major issue facing businesses around the globe today.

2.9 Summary

This chapter described the overall look, process, benefits, vendors, challenges and future of IIoT. Following this, the chapter discussed the differences between IIoT and IoT. Then we outlined the application areas of IIoT in detail. Industrial IoT has the potential to change production completely. As the world market and industry power push producers to consider performance, intelligent production running by IIoT data driven by information analysis is necessary. The fact is, numerous data will generate from highly interconnected machine systems, and the understanding of important and valuable functional insights thought IIoT is unlimited. IoT is changing recent industry into intelligent industry. But, the core aspects of the IIoT lead to many challenges, for example, decentralization, poor mobility, and safety and confidentiality risks. Blockchain technique offers chances to address IoT challenges. Therefore, the next chapter discusses blockchain technology in detail.

References

[1] Karmakar A, Dey N, Baral T, Chowdhury M and Rehan M 2019 Industrial Internet of Things: a review *IEEE 2019 Int. Conf. on Opto Electronics and Applied Optics*

[2] Tomas J P 2017 Thames water rolls out smart meter project in london https://wiprodigital.com/cases/progressive-metering-a-utilitys-strategic-move-into-predictive-planning/

[3] Daugherty P, Banerjee P, Negm W and Alter A E 2015 Driving unconventional growth through the Industrial Internet of Things, Accenture Technology

[4] Yu X and Guo H 2019 A survey on IIoT security *2019 IEEE VTS Asia Pacific Wireless Communications Symp.*

[5] Sisinni E, Saifullah A, Han S, Jennehag U and Gidlund M 2018 Industrial Internet of Things: challenges, opportunities, and directions *IEEE Trans. Industr. Inform.* **14** 4724–34

[6] Anis M, Gadallah Y and Elhennawy H 2016 Machine-to-machine communications over the Internet of Things: private Wi-Fi access prospects *IEEE 2016 Wireless Days (WD)*

[7] Buntz 2017 The top 20 industrial IoT applications https://iotworldtoday.com/2017/09/20/top-20-industrial-iot-applications/

[8] Yang C, Shen W and Wang X 2016 Applications of Internet of Things in manufacturing *2016 IEEE 20th Int. Conf. on Computer Supported Cooperative Work in Design*

[9] Breivold H P and Sandström K 2015 Internet of Things for industrial automation—challenges and technical solutions *2015 IEEE Int. Conf. on Data Science and Data Intensive Systems*

[10] Jahan S, Rahman M S and Saha S 2017 Application specific tunneling protocol selection for virtual private networks *2017 IEEE Int. Conf. on Networking, Systems and Security*

[11] Fraile F, Tagawa T, Poler R and Ortiz A 2018 Trustworthy industrial IoT gateways for interoperability platforms and ecosystems *IEEE Internet Things J.* **5** 4506–14

[12] Wollschlaeger M, Sauter T and Jasperneite J 2017 The future of industrial communication: automation networks in the era of the Internet of Things and Industry 4.0 *IEEE Ind. Electron. Mag.* **11** 17–27

Lakshmana Kumar Ramasamy and Seifedine Kadry

Chapter 3

Blockchain technology

This chapter presents fundamental and improved blockchain notions. Blockchain is a continuously-growing ledger that maintains a constant record of all business dealings in a safe, period sequence and unchanging manner. It can be utilized for the safe exchange of cash, assets, agreements and so on. No third party mediator like the government or a bank is needed. Blockchain is protocol software, but it cannot operate in the absence of the internet. This chapter debates blockchain's overview, its structure, and how it is executed in different aspects.

3.1 Overview of blockchain

Before exploring blockchain, it is significant to understand why this latest technology appeared. A response was required to the danger of called double-spending [1]. Note the scenario demonstrated in figure 3.1.

A figure 3.1 show, Alice pays Bob $10 to buy some books. Once Bob gets this $10, Alice has no way to use the money again for any other dealings because the money is at present in Bob's ownership. At present, note the scenario is where payments are made in digital form. It is explained in figure 3.2.

Since the format for the currency transfer is in digital form, it is a physical binary file saved somewhere on Alice's computer. After Alice presents this file (digital currency) to Bob, she can present a copy of it to Lisa too. They both believe that they have earned the currency without any way of recognizing the digital currency and thus deliver the respective merchandise to Alice. It is known as double-spending, where the buyer pays the same amount of money to get services or merchandise from numerous merchants in many locations. To resolve this issue of double-spending, one can hire a centralized authority to supervise all transactions. It is demonstrated in figure 3.3.

The centralized authority, usually the bank, keeps a ledger mantaining a record of all business dealings. At present, Alice has to transmit her digital currency to the bank, which enters her ledger that debits Alice's account. After confirming that Alice has enough money to pay for the digital currency she requires to transmit, it will

doi:10.1088/978-0-7503-3663-5ch3

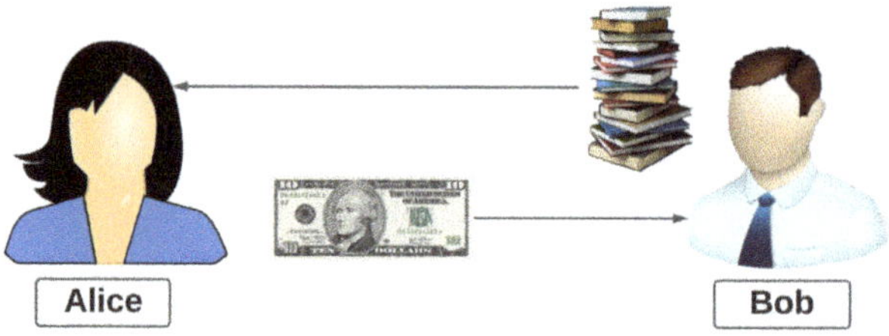

Figure 3.1. Alice pays Bob $10 to buy some books.

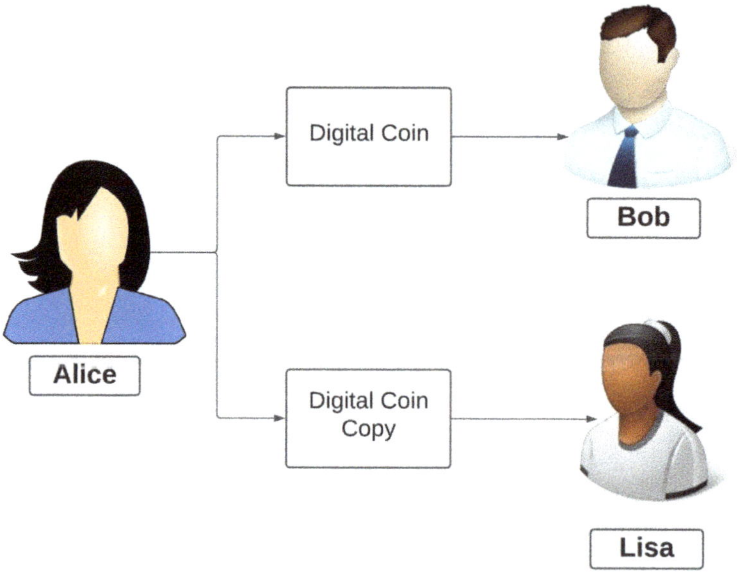

Figure 3.2. Currency transfer is in digital format.

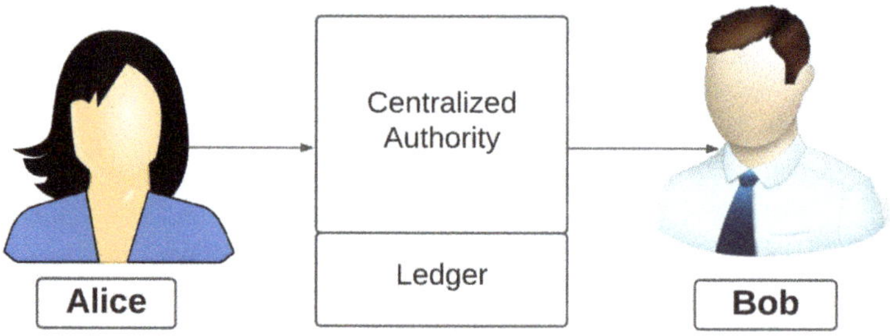

Figure 3.3. Hire a centralized authority to supervise all business dealings.

transmit currency to Bob from the credit in her account in its ledger. At present, it has confirmed that Alice cannot spend money twice. If all digital transactions are transferred to a centralized authority like this, the issue of double-spending can be resolved. It presents one further advantage in ensuring the authenticity of each coin (digital currency) that it receives in business dealings. So counterfeit currency (duplicate currency as happened to Alice who pays Lisa using a copy) will be discovered and prevented from streaming.

Although the launching of centralized authorities resolved the double-spending issue, it also launched one more important issue—the price of constructing and maintaining the centralized authority. As banks required money to carry out their duties, they began to deduct commissions on each transaction they made for their customers. It can occasionally be quite costly, in particular in foreign transfers where many agents (banks) can be involved in the whole agreement. All of the above issues can be resolved by introducing a digital currency, namely bitcoin. This section provides a short description of what bitcoin is before exploring in depth its plan and structure.

Bitcoin was launched to the globe in 2008 by Satoshi Nakamoto. Not only did bitcoin solve the double-spending issue, but it further provided numerous other benefits, one of which is the anonymity of transactions. Satoshi who generated and traded some coins in this system is utterly anonymous to the whole world.

As you have seen before, the bank keeps a ledger that records all business dealings. This ledger is kept private and maintained by the bank. Satoshi suggested that the ledger should be public and owned by society. In the meantime, when you create such a ledger in public, you can think of it many thoughts. This ledger must be tamper-proof so that no one can alter its entries. Since all entries of this ledger are openly viewable, it must be kept anonymous; you wouldn't want everyone in the globe to recognize that somebody paid you $10. Also, since there is only one ledger to keep track of all global business dealings, the ledger size can represent a further major anxiety. It is not trivial to provide a resolution to these problems; this chapter is intended to show the basic structure of bitcoin in simple terms. To know the structure of blockchain, you require knowing some of the key components on which it based. So, let us start with public-key cryptography (PKI).

3.2 Public key cryptography (PKI) in the blockchain

PKI is also called asymmetric cryptography [2]. It utilizes two keys—private and public. The key is a long binary number. The public key is shared throughout the world and is public as its name implies. The private key must be private and never be lost. In the case of bitcoin, if you lose the private key in your wallet of bitcoin, all your wallets' contents will be in danger of being stolen immediately; there will be no evidence of who stole all of your money (in your wallet)—that is the anonymity mentioned at the beginning of this chapter.

PKI executes two operations—to verify the authenticity and privacy of the message using encryption/decryption. This section will describe both of these operations.

3.2.1 Authentication

When exchanging messages with both parties, it is important to build reliability between the sender and the recipient. In particular, the recipient should trust the source of the message. Going to our previous scenario (shown in figure 3.1) of Alice transferring money to Bob by buying merchandise from him, let us see how PKI constructs this reliability between Alice and Bob. See figure 3.4.

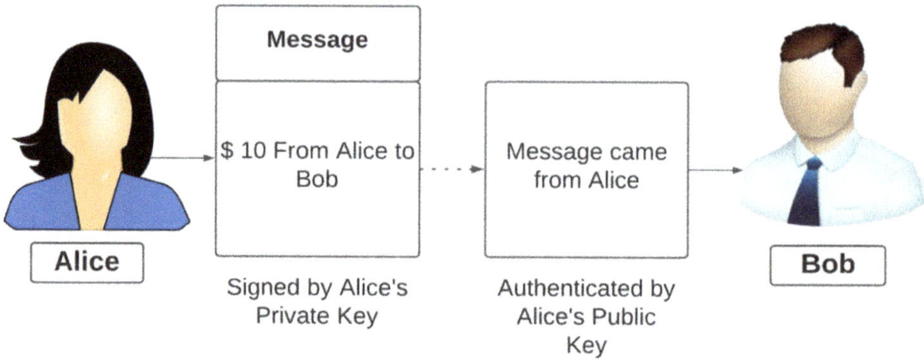

Figure 3.4. PKI constructs reliability between Alice and Bob.

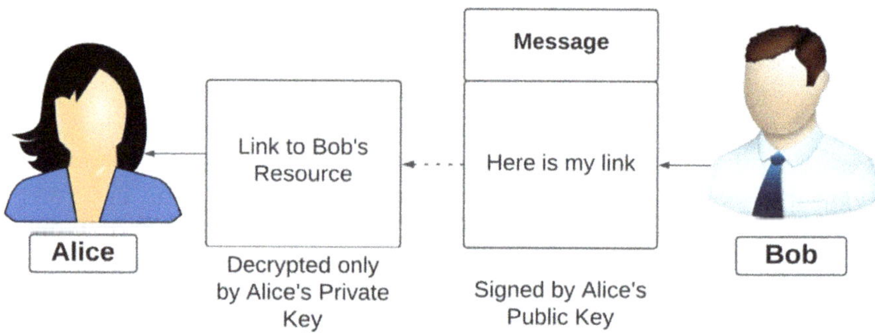

Figure 3.5. Bob generates a message and transmits it to Alice.

First, if Alice desires to transmit money to Bob, she has to generate her public and private key. Consider that two keys are at all times occurring in pairs jointly, and you cannot combine public and private keys for various people or situations. At present, Alice says she sends $10 to Bob. So she generates a message (a plain text message) having Alice's public key (sender), Bob's public key, and the sum ($10). The reason for this money transfer, for example, 'I desire to purchase some books from you' is further contained in this message. Every message is currently signed using the private key of Alice. When Bob acquires this message, he will utilize the PKI signature validation algorithm and the public key of Alice to verify that the message came from Alice. It launches the reliability of the message creator. Next, let us take a look at the privacy of the message.

3.2.2 Privacy of the message

At the time Bob has obtained his amount paid, he desires to transmit a link to his ebook that Alice desires to purchase. So Bob will compose a message he transmits to Alice, as demonstrated in figure 3.5.

Bob generates a message, for example 'Here is the link to the book you requested', signs it with the public key of Alice (obtained from the request message of Alice) and encrypts the message using the secret key distributed between the two during the handshake of the HTTPS transfer. Now Bob knows that only Alice can decipher the message utilizing the private key held by Alice. Also, someone obstructing the message will not be capable of retrieving its contents for the reason that the contents are encrypted by a secret key held solely by Alice and Bob. It confirms to Bob that use to his ebook is only given to Alice.

3.3 Hashing

One of the primary techniques in PKI is a hash technique [3]. The hash technique maps any arbitrary amount of information to static size information. Bitcoin utilizes the SHA-256 hash technique to generate a hash of 256 bits (32 bytes). This is explained in figure 3.6.

When Alice places an order with Bob, she generates the same message as the one displayed above. This message has been hashed with a hash technique that generates a hash of 32 bytes. The usage of this hash is for entirely practical reasons, the hash (256 bit value) regarded as distinctive in the message content. Once the message is changed, the value of the hash will alter. Not only does the hash provide value, but the original message also cannot be rebuilt. Afterwards looking at the significance of hashing, we will look at one more concept, which is mining in bitcoin.

3.4 Bitcoin mining

While Alice generates a purchase requisition for Bob, she does not transmit it to Bob only. The purchase requisition is transmitting over the whole network to which she is linked [4]. The network of Alice is shown in figure 3.7.

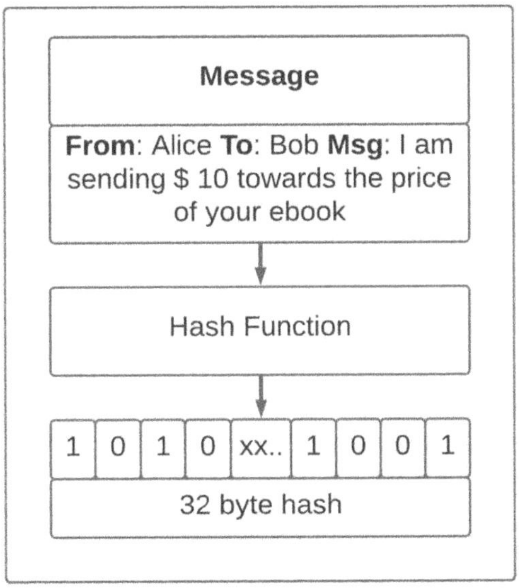

Figure 3.6. Hashing.

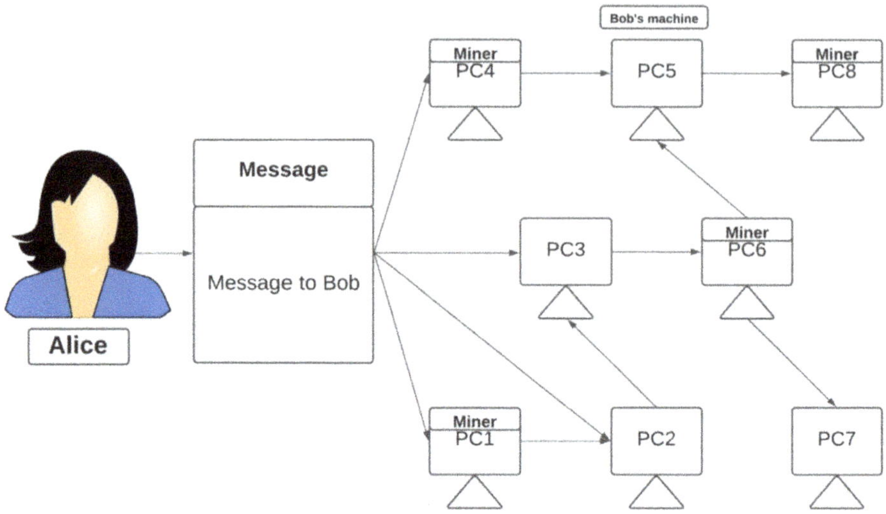

Figure 3.7. The network of Alice.

The purchase requisition goes to all linked machines. A few nodes on the figure are denoted as miners. These are machines that use software to mine the message of bitcoin.

3.4.1 Process of mining

While the whole network spread is over a wide area, it is anticipated that every miner in the network would attain numerous messages from numerous sellers at any provided time. The miner's role is to merge these messages into a single block. It is explained in figure 3.8.

After the message block is built, the miner makes a block hash using the hashing technique as previously explained. Therefore, one knows that if a third party is to change the content in this block, its value of hash will be incorrect. By chance, each message is time-stamped so that noone can change its order without changing the hash value of the block. Therefore, messages in the block are completely protected from intrusion. How this fact is utilized in the protection of entire network dealings is also described.

3.5 Chaining blocks

Blocks made by different miners are grouped jointly and called a shared public ledger [5]. It is explained in figure 3.9.

Every block in the chain has numerous messages (dealings), as seen previously in figure 3.8. A block can arrive from any miner in a chain. When generating the sequence of blocks, we note the rule that the hash of the preceding block is attached to the present block. So, the miner, when making the block, takes the hash of the preceding block in the network, attaches it to its message set and makes a hash of its recently generated block. This recently constructed block is changed into the new

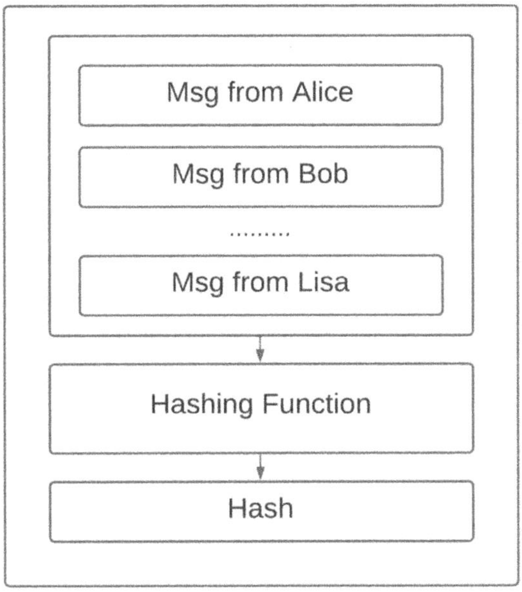

Figure 3.8. Miner merges a lot of messages into one block.

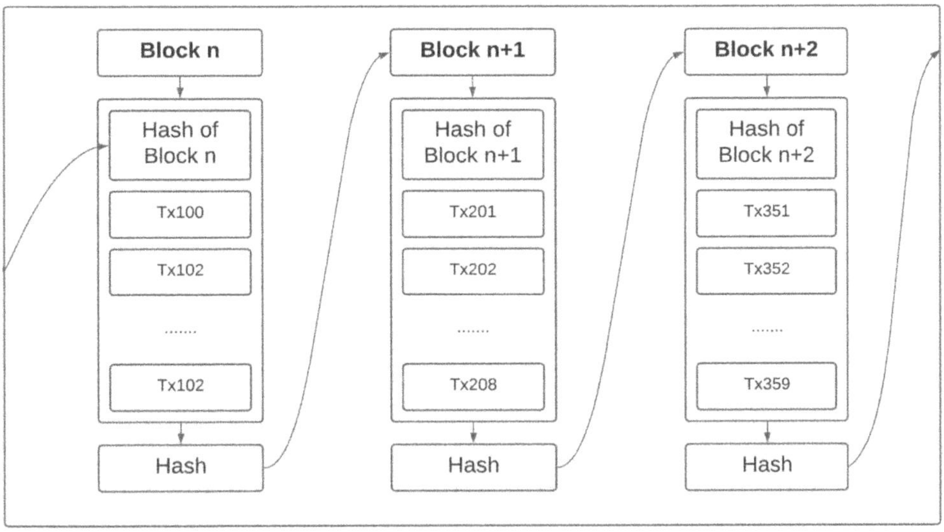

Figure 3.9. Blocks chaining.

end of the chain, and thus the chain continues to grow as further blocks are attached to it by miners.

3.6 Proof of work (PoW)

Since entire transactions are time-stamped, a shared time-stamp server in the peer network needs to be used [6]. It needs further execution, and that is the PoW that

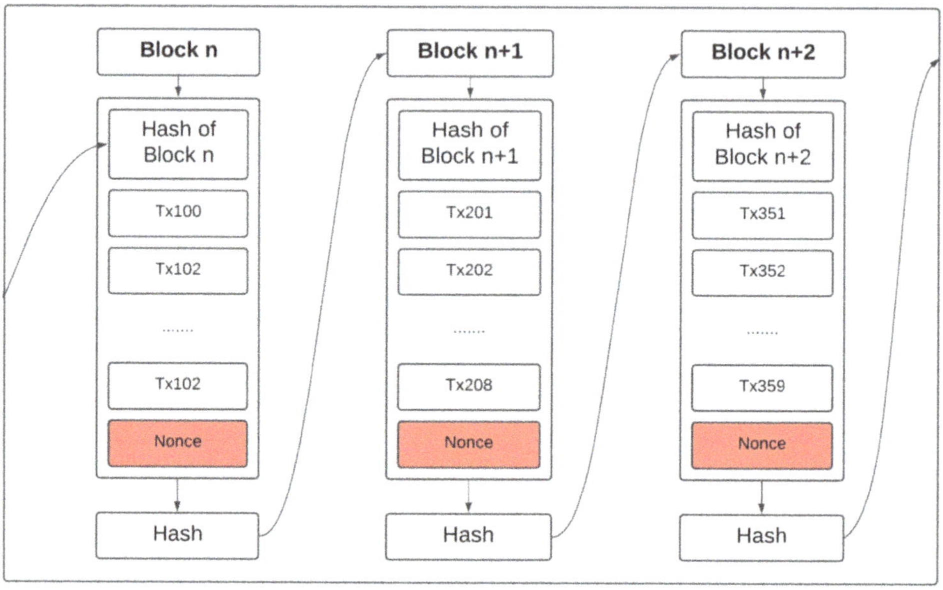

Figure 3.10. Blockchain PoW.

this section explains. For every block, we attach another item named Nonce, as demonstrated in figure 3.10.

The Nonce is some value in the same way that the hash value of the block meets specific criteria. The hash created by these criteria should contain its foremost four-digits as zero. Therefore, the created hash will look similar to 000011101110xxx. Normally, the miner begins with a nonce value of zero and increases the hash created up to it meeting specific criteria. Consider that the hash creation runs randomly and is uncontrollable. Therefore, it can get numerous repetitions as long as the required hash is produced with four leading zeros. The anticipated time in a bitcoin system to create a block is 600 s. When the miner has effectively mined the block, the system then creates the final block of the chain.

Many miners are contesting to produce an official block. The Bitcoin scheme rewards the first victorious miner with a few Bitcoins. Generally, a powerful miner can be the first winner. It can lead to attacks on the whole system by those with large processing power. The final sections will explain forms of attack and how to reduce them.

3.7 Bitcoin mining and network

This section gives a brief statement of the main points of the steps outlined above [7]:

1. Anyone who desires to receive goods from a third party seller first makes a transaction (message to the selected receiver).
2. In the allotted time, there may be multiple senders (buyers) and recipients (sellers) who generate the transaction.

3. Each transaction is transmitting to the network across all nodes. The transaction provided does not have to be attained somewhere in the network.
4. Every node gathers the latest transactions into a block. Consider that the collection of business dealings in every block is self-sufficient of the blocks collection generated by others and is in a natural manner dissimilar from others. It does not matter; the system guarantees that all transactions transmitted on the network are contained in a few blocks for a limited time. Typically, the sender will promote this node by providing a specific cost of bitcoins to the miner for its endeavours. The miner can choose to give preference to those with more offers.
5. The node is then working on discovering a PoW for its assembled block.
6. If the node discovers the PoW, it transmits the assembled block to the network.
7. The nodes that obtain the latest block would agree with it only after checking that all business dealings in the block are accurate and not before they are completed.
8. If a block is agreed as precise, the node operating in its latest block would need to reconnect the transaction to its block to ensure that the transaction is not copied. The node then works to get the PoW on its recently built block; when performing this, it would get the agreed block hash as the preceding hash.
9. Similarly, the blockchain persists in growing indefinitely.

3.8 Incentives to miners

In bitcoin mining, a miner can be flooded with more transactions at any time [8]. Most block sizes are predefined in a system that requires a specific transaction number to enter a block. The value of block inputs is decided by the size of the predefined block and the mean length of every block. Important advice here is that the sender should not add too much data to their message, thus making it shorter, thus encouraging the miner to acknowledge it before other long messages. The sender will usually further include a transaction cost based on a specific number of bitcoins, thus encouraging the miner to add to his block at the beginning. Another result of constructing a blockchain is just its size. Over time, the whole blockchain can become excessively large for the node to maintain on its disk space. It is resolved based on the Merkle tree explained below.

3.9 Merkle tree of blockchain

The problem of drive space at one node can be simply conquered when all business dealings of a block are hashed on a Merkle tree [9], as displayed in figure 3.11.

Currently, the header of the block has the hash of the previous block, the root hash of the complete business transactions on the recent block, and a Nonce in a Merkle tree. Since this root hash incorporates hashes of all business dealings inside the block, these business dealings can trim to free up disk space. The resulting blockchain would appear as in figure 3.12.

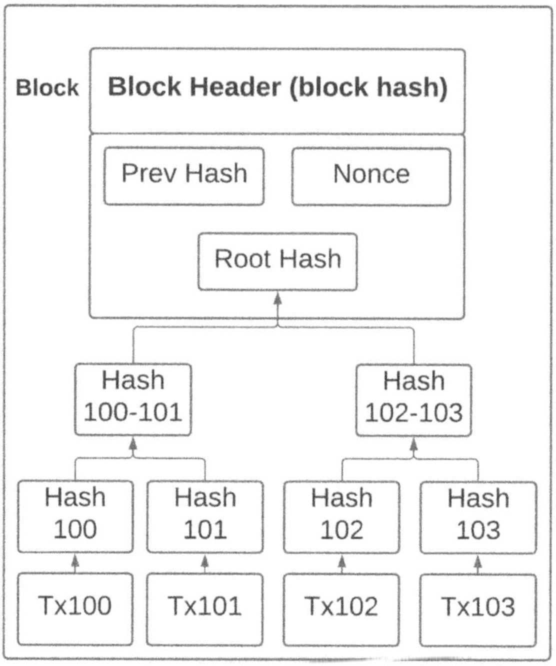

Figure 3.11. Transactions hashed in a Merkle tree.

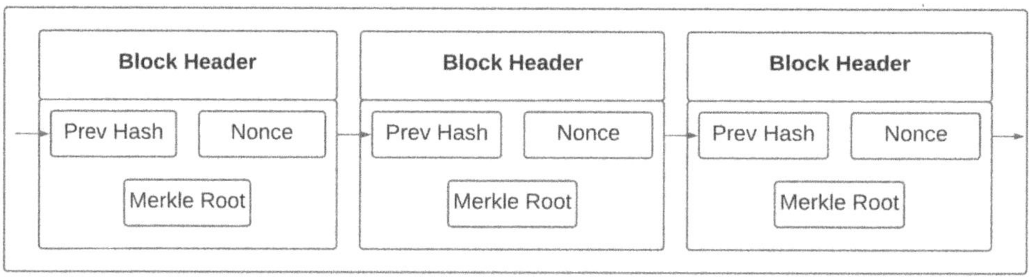

Figure 3.12. Blockchain architecture.

It can lead to significant reductions in drive space usage. This plan is utilized by a regular customer who is concerned with obtaining amounts from others. But, miners require maintaining a complete blockchain. The question then arises, how does a recipient check the fee in the absence of the capability to locate the obtained currency correct to its beginning. This is described below.

3.10 Payment verification in blockchain

Consider an instance where you want to check a specific fee you made earlier as a seller. The blockchain you have on your computer has only the block headers as demonstrated in figure 3.12, so the transaction you are looking for is not in your blockchain copy. You could at present look back at a copy of your blockchain to discover a block stamped with the desired transaction time. Then you would consult the Merkle tree of the chosen block, and you would retrieve the transaction you want [10]. This is explained in figure 3.13.

If your target is Tx103, even if you cannot observe the contents of Tx103, you do understand which block it belongs to and all corresponding following blocks in the chain. So you can trust the security of this transaction and continue with your trade.

3.11 Conflicts resolution

As mentioned above, there are many miners in the network of Bitcoin. It is feasible that two separate miners will simultaneously resolve the PoW, therefore adding their blocks to the final well-known block in the blockchain [11]. This is explained in figure 3.14.

Currently, there are two branches after block 103. Both of these branches are precise. Therefore, a subsequently mined block can attach to both branches. Assume a miner attaches a recently mined block to Block 104-A, a branch accompanied by

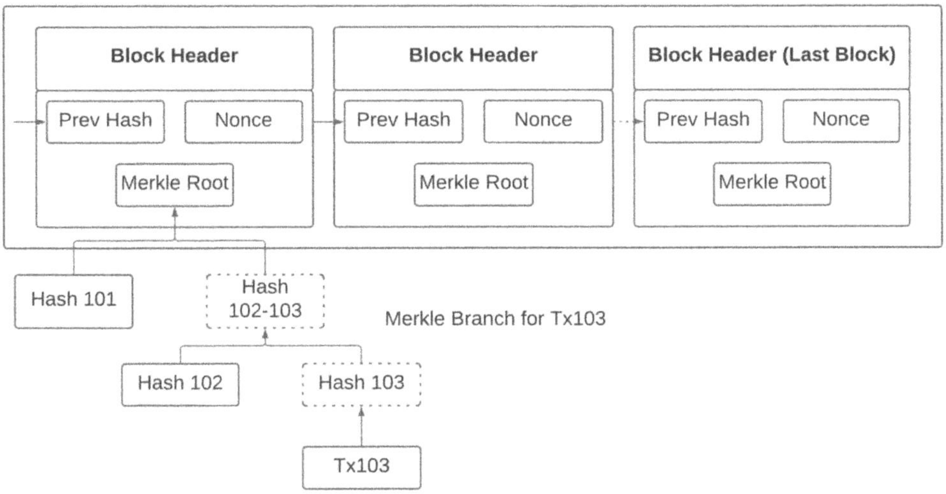

Figure 3.13. Payment verification in blockchain.

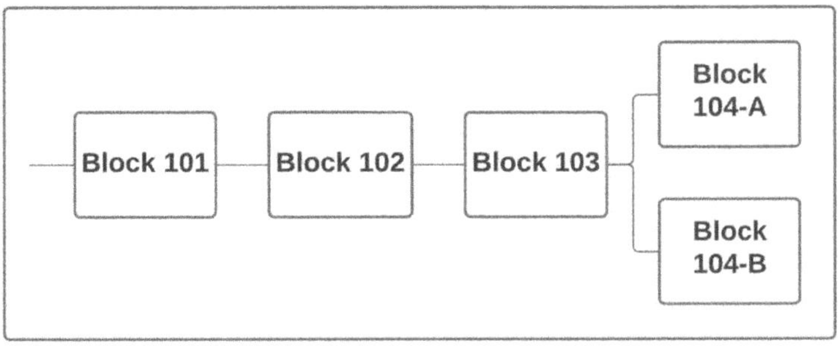

Figure 3.14. Two separate miners have resolved the PoW simultaneously and therefore added their blocks to the final well-known block in the blockchain.

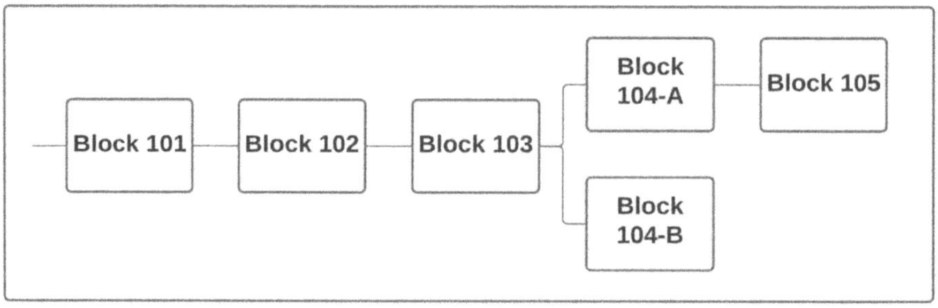

Figure 3.15. The length branch is at all times successful, and the shortest branch is eliminated.

Block 104-A would be longer than a branch accompanied by Block 104-B, as shown in figure 3.15.

In the structure of bitcoin, the longest branch at all times is successful and shorter ones are eliminated. Therefore, Block 104-B should be eliminated. Before removing this block, each transaction in this block would go back to the pool of transactions; here they can be obtained a mine and attached to the upcoming block.

3.12 Privacy in blockchain

Privacy is at risk with regard to disclosure of the ledger that records all Bitcoin transactions [12]. Anyone on Earth could then recognize who is paying who. The conventional financial institution can keep this type of privacy by maintaining its records in secret. In the Bitcoin network, privacy is attained through a different plan. Recall we said that the Bitcoin sender needs to understand who to pay. Therefore, they request the public key of the seller. This public key is unknown. In other words, as a service provider, if someone asks you where to transmit money, simply send them your public key. The combination of this public key is not recorded any place in this ledger. That way, anyone external to this transaction will only understand how much is paid to and how much received by the public key. To attain the highest level of privacy, you may generate the latest public/private key for every transaction; therefore, most of your transactions can be collected by a third party. To an outsider, this may mean that most small transactions are created and cannot be connected to the same source.

Finally, any internet-based program is at risk of misuse. The following section explains some kinds of attacks on the bitcoin system and how to reduce that.

3.13 Mitigating attacks on bitcoin

This section will discuss three different kinds of possible attacks on bitcoin [13].

3.13.1 Race attack

The attacker can quickly transmit similar bitcoins to different merchants, perhaps based on two different computers. If sellers do not wait for the confirmation of the blocks before providing the merchandise, they would not know that a transaction was discarded in the process of mining. The resolution to this type of attack is for the seller to wait at the minimum for a block verification before shipping the merchandise.

3.13.2 51% attack

In this type of attack, we get the impression that someone owns 51% of the network's computing power. In these types of attack mines, the attacker mines a private blockchain where he pays out the same coin twice. Since he holds most of the power for computing, he is assured that in a few cases his block would be longer than the chain of the 'trusted' network. He then publishes his blockchain into the system, invalidating all transactions previously stored on a trusted blockchain. This type of attack is deceptive because acquiring power for computing is more costly, equivalent to or greater than 51% of the power for computing of the whole network.

3.13.3 Finney attack

In this instance, an attacker is a miner. A miner makes a block with his transaction and cannot publish it on a system. He then utilizes similar coins in the second transaction and then issues the mined block. The second transaction will finally be discarded through further miners; however, this would take a while. To minimize this danger, the vendor must stay for not less than six blocks to secure the goods before shipping.

3.14 Conclusion

This chapter discussed many blockchain concepts by using bitcoin as an example. Bitcoin is the initial victorious execution of blockchain. Nowadays, the world has discovered blockchain applications in many industries, where confidence is required in the absence of the participation of a centralized authority. The next chapter discusses the key features of blockchain technology.

References

[1] Zheng Z, Xie S, Dai H, Chen X and Wang H 2017 An overview of blockchain technology: architecture, consensus, and future trends *2017 IEEE Int. Congress on Big Data*

[2] Osmov V, Kurbanniyazov A and Hussain R 2019 On the blockchain-based general-purpose public key infrastructure *2019 IEEE/ACS 16th Int. Conf. on Computer Systems and Applications*

[3] Bhulania I P and Raj G 2018 Analysis of cryptographic hash in blockchain for bitcoin mining process *2018 IEEE Int. Conf. on Advances in Computing and Communication Engineering*

[4] Kaur Chadha G and Singh A 2019 Bitcoin block-chain mining *2019 IEEE 9th Int. Conf. on Cloud Computing*

[5] Liu S and He S 2019 Application of block chaining technology in finance and accounting field *2019 IEEE Int. Conf. on Intelligent Transportation, Big Data & Smart City*

[6] Gemeliarana I G A K and Sari R F 2018 Evaluation of Proof of Work (POW) blockchains security network on selfish mining *2018 IEEE Int. Seminar on Research of Information Technology and Intelligent Systems*

[7] Zhu J, Liu P and He L 2017 Mining information on Bitcoin network data *2017 IEEE Int. Conf. on Internet of Things (iThings)*

[8] Chang S Y and Park Y 2019 Silent timestamping for blockchain mining pool security *2019 IEEE Int. Conf. on Computing, Networking and Communications*

[9] Sharma B, Sekharan C N and Zuo F 2018 Merkle-tree based approach for ensuring integrity of electronic medical records *2018 9th IEEE Annual Ubiquitous Computing, Electronics & Mobile Communication Conf.*

[10] Pouraghily A and Wolf T 2019 A lightweight payment verification protocol for blockchain trans. on IoT devices *2019 IEEE Int. Conf. on Computing, Networking and Communications*

[11] Destefanis G, Marchesi M and Ortu M 2018 Smart contracts vulnerabilities: a call for blockchain software engineering *2018 IEEE Int. Workshop on Blockchain Oriented Software Engineering*

[12] Bernabe J B, Canovas J L and Hernandez-Ramos J L 2019 Privacy-preserving solutions for blockchain: review and challenges *IEEE Access* **7** 164908–40

[13] Shalini S and Santhi H 2019 A survey on various attacks in bitcoin and cryptocurrency *2019 IEEE Int. Conf. on Communication and Signal Processing*

Lakshmana Kumar Ramasamy and Seifedine Kadry

Chapter 4

Blockchain key characteristics

Blockchain technology has been around for some time, attracting ever more serious attention. This technology originally attracted attention to the very famous cryptocurrency bitcoin. Blockchain technology is not only a backup network for cryptocurrencies, but it also provides much more. So what are the critical blockchain characteristics that make it inevitable? Why is it so famous? To reply to these questions, we will look a little deeper into the features of blockchain in this chapter. Figure 4.1 shows six essential blockchain key characteristics.

4.1 Immutability

There are a few exciting blockchain characteristics; however, their 'immutability' is doubtless one of the critical attributes of blockchain technology. But why is this technology so erratic? Let us begin with the blockchain that connects, with immutability.

Immutability is something that cannot be modified. Being a constant, unchangeable network is one of the best blockchain characteristics to keep the technology intact. Blockchain technology operates a little differently than a traditional financial institution. Rather than trusting on centralized authorities, it guarantees blockchain characteristics via a set of nodes [1].

Each node of the network contains a duplicate of the digital ledger. Each node must verify its validity to attach a transaction. If the majority believes it is valid, it will be included in the ledger. This encourages lucidity and prevents fraudulence. Therefore, no transaction module could add to the ledger in the absence of the consent of the greater number of nodes.

Another reality is that backing up the record of crucial blockchain characteristics means that once the transaction blocks are attached to the ledger, nobody could go back and modify it. As a result, no user on the network could change, remove, or update it.

doi:10.1088/978-0-7503-3663-5ch4

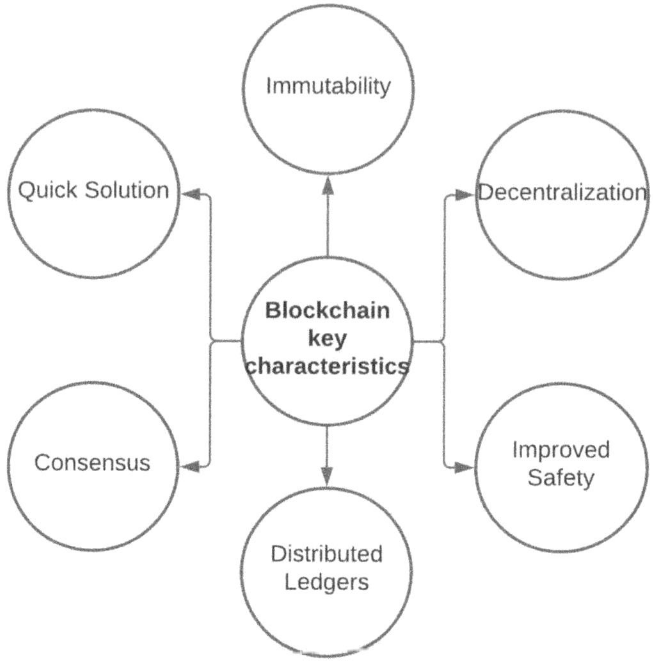

Figure 4.1. Essential blockchain key characteristics.

4.1.1 How does it confront fraudulence?

It is evident that a lot of money is hacked by normal channels yearly. Humans spend trillions of dollars to secure their business from any outside hacks. But, we often neglect to calculate the inside cybersecurity threats, which occur from corrupt individuals and authorities.

Frequently, these hacks have an inside connection learning the safety measures; therefore, eventually, we pay the cost of our confidence. It could be argued that banks are not credible at present and the world economy requires believable surroundings to tackle this problem thoroughly. Therefore, because it is free of the surrounding corruption, one could presume that blockchain will certainly greatly alter these specific possibilities. Once businesses begin to coordinate blockchain technology to maintain their internal networking schemes, nobody will be capable of hacking or changing or stealing data.

Public blockchains are an excellent instance of this. Everybody in the public blockchain could view the transactions; therefore, it is self-evident. Conversely, a federal or private blockchain is ideal for companies that require their data to be transparent among employees but shielded from public vision.

4.1.2 How is immutability achieved?

The hash value protects each block of the code individually. However, it is interesting to see how this sets up immutability. To understand this situation one needs to comprehend cryptographic hashing.

Figure 4.2. Immutability.

Nowadays, creating a cryptographic hash is not daunting because current programming languages offer many 'hash functions'. Herewith, one merely has to send a byte, and the method would provide a checksum signature. There is no shortage of ways under this umbrella, but SHA-256 is famous in the world of blockchain. Let us explore this with an example.

You need to create a cryptographic hash in Python code. For this, you need to import the hashlib package from Python's standard library that provides access to the sha-256 method. These methods form a string of 64 characters. Regardless of the input size, you would obtain every time the similar stable length of the line as the digital signature.

This digital signature indicates the precise information you would be entering. The leading utility of this hash is that you cannot modify it. It means you cannot utilize this result string to discover input information. It merely causes immutability [2]. Figure 4.2 explains immutability.

Transactions proved by a blockchain network comprise volumes of data implanted with time-stamps that are protected by a hashing procedure. It connects jointly and connects the hash of the preceding block. This algorithm creates a sequential chain which secures each block.

Hashing every time incorporates the metadata of the preceding block, for which a novel hash is created that sets up a connection among the block and the chain, which are then 'non-breakable'. Thereafter, nobody could remove and change the information of the block situated in the blockchain, since if someone tries to do this, the next block will reject the change (since the block hash is no longer valid). It is correct that this is a powerful method. But, there are many challenges overcome in this method.

4.1.3 Challenges to Immutability of Blockchain

51 per cent attack

There are many challenges to this method. But, the main frailty may be the chance of a '51 per cent attack' [3]. The word implies that an attacker could gain enormous calculating energy above the entire other members of the network. In short, this could be described as 'managing the urge to create energy'.

It is a decentralized structure of a network where no company is responsible. Nevertheless, miners jointly could cause the demise of the immutability of the blockchain method. At present, due to the rise in mining markets and access to rental mining volume, a human has no complication in carrying out such an attack. It permits hackers to change transaction information that was initially considered 'unchanged'. With this feature, hackers could convert large value transactions, expend cash a second time and make an income.

Quantum computing
Another major challenge of this space is quantum computing, which threatens the immutability of the blockchain. IBM specialists emphasize that quantum computing can change the public key of the blockchain network that could detect private keys to infiltrate the method. There is no uncertainty that this is an actual and acceptable danger to the space, which can infect nearly 50 per cent of blockchains.

4.1.4 Solution

Experts recommend that the '51 per cent attack' could be overcome by making a robust protocol and utilizing a consensus algorithm, for example, the assigned proof-of-stake or exactly 'proof-of-stake' algorithm. It is more difficult to place the number of tokens on the network than hiring computing energy. However, it is hard to say whether these solutions are sufficiently dependable in the face of such a menace.

For the menace of quantum computing, more application developers have suggested integrating it within the centre of quantum cryptography blockchain. In the years to come, the blockchain structure generated by quantum particles would be capable of storing entire history in a very safe method. Solutions for these are very modernistic. Therefore, for the moment, one must be careful in accepting blockchain and the advantages that are obtainable to build healthy application resolutions.

4.2 Decentralization

The network is decentralized, with no administrative authority or a person overseeing the structure. The collection of nodes expands the network. It is one of the critical characteristics of blockchain technology. Blockchain puts consumers in an uncomplicated location [4]. Since the technology does not need any administrative authority, one could access it straight from the internet and save our assets there.

One could save everything from cryptocurrencies and significant contracts to documents or other beneficial digital assets. With the assistance of blockchain, you have direct control over them based on your private key. Therefore, you understand the decentralized structure returns the energy and rights of their property to the public [5].

4.2.1 Why is this so useful?

Now let us view how this blockchain characteristic creates a difference, as shown in figure 4.3.

- **A smaller amount of failure:** All things in the blockchain are completely structured, and it is hugely fault tolerant as it does not rely on manual computation. Therefore, unintentional failures of this scheme are not a common issue.

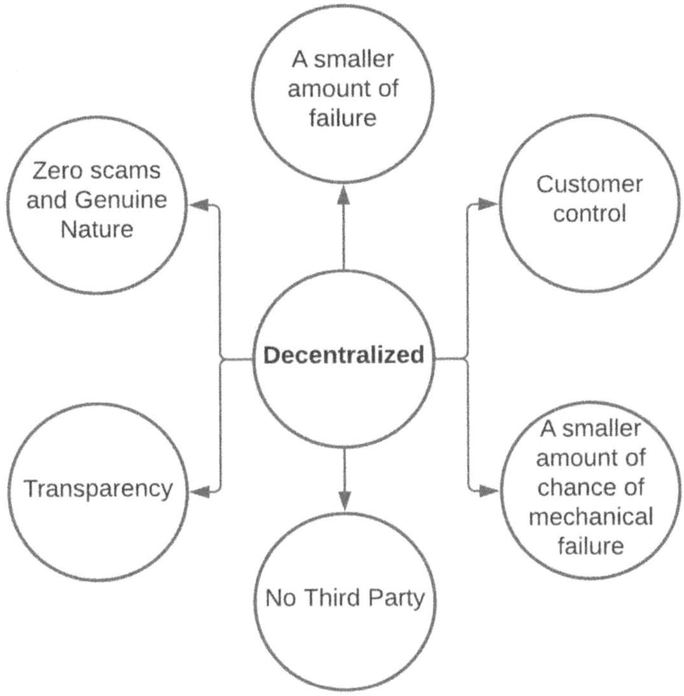

Figure 4.3. Reason why the decentralized nature of blockchain is so useful.

- **Customer control:** With the deployment, customers at present maintain control over their possessions. They do not have to trust any third party to keep their assets. They could all do it at once.
- **A smaller amount of chance of mechanical failure:** Decentralization is one of the critical characteristics of blockchain technology that could escape any malicious attack. This is because attacking a scheme is a costly and challenging solution for hackers. Therefore, it is less likely to cause a mechanical failure.
- **No third party:** The decentralized character of technology makes third-party industries an unreliable organization; no third party, no additional risk.
- **Transparency:** The decentralized character of technology creates an apparent profile of each contributor. Every change in the blockchain is visible, which makes it very stable [6].
- **Zero scams:** Because the scheme is run on mechanisms, humans are less likely to cheat you on anything. Nobody could use blockchain for their gain.
- **Genuine nature:** This nature of the organization creates a distinctive organization for each type of human. Attackers would find it hard to destroy.

4.3 Improved safety

Blockchain eliminates a central authority so that nobody could alter any of the properties of the blockchain for their advantage. The use of encryption guarantees one more layer of scheme safety. How does it provide so much protection compared to previous technologies? Well, it is very safe since it gives a distinctive enhancement —cryptography [7]. With the decentralization, cryptography adds one more layer of security for customers. Cryptography is an extremely mixed mathematical method that works as a firewall for attacks [8].

Every piece of data in the blockchain is cryptographically hashed. In simple terms, the data in the network obscures the genuine nature of the information. For this procedure, any input information generates various values by a mathematical method, but the length is adjusted every time.

One can trust this as a distinctive identifier for each data. All volumes in the ledger come with their unique hash and contain the hash of the last block. Therefore, altering or attempting to damage data is equivalent to modifying every hash ID. That is unfeasible. You possess a private key to use the information; however, there would be a public key to create transactions.

4.3.1 Hashing is irreversible

Hashing is so composite that it is unfeasible to alter it. Nobody could get the public key and combine it with a private key [9]. Furthermore, change in input can lead to an entirely different ID, therefore small variations are not accommodated on the scheme. When somebody needs to disrupt the network, he must replace all data saved on each node of the system.

There may be millions and millions of humans, where everybody has a similar duplicate of the ledger. Using and hacking millions of systems is unfeasible and expensive. That is why it is one of the better blockchain characteristics. Because it is thus difficult to ignore, you do not have to be concerned about hackers getting all your digital benefits from you.

4.4 Distributed ledgers

Generally, a public ledger would offer all data regarding a transaction and contributors. It is all unlocked, nothing hidden. However, the instances of private or federal blockchain are slightly dissimilar; in those incidents, more humans could view what is going on in the ledger.

The ledger in the network is maintained by all other customers on the scheme. It shares computing energy across systems to guarantee the best result. This is regarded as one of the critical attributes of blockchain. The outcome would always be a high-efficiency ledger scheme that could pick up conventional ones.

The good news is that despite acronyms like DLT (distributed ledger technology) in fintech circles, this technology is comparatively simple to know. A distributed ledger is a database that works in multiple positions or between numerous contributors. In contrast, many industries presently utilize a centralized database that exists in a stable place. A centralized database contains a single point of failure (SPOF) [10]. Figure 4.4 shows a centralized ledger.

But, a distributed ledger is disseminated to remove requirements for a central authority or mediator to execute, verify or authorize transactions [11]. Companies utilize a distributed ledger method to implement, verify, or confirm transactions or

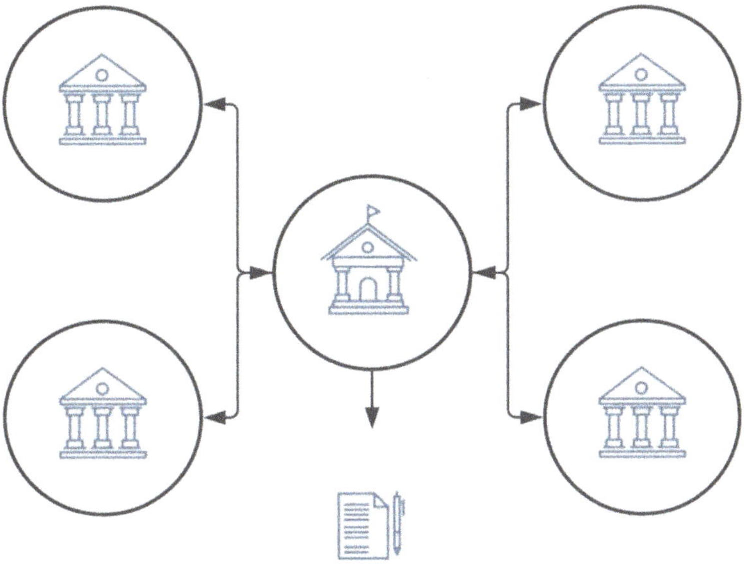

Figure 4.4. Centralized ledger.

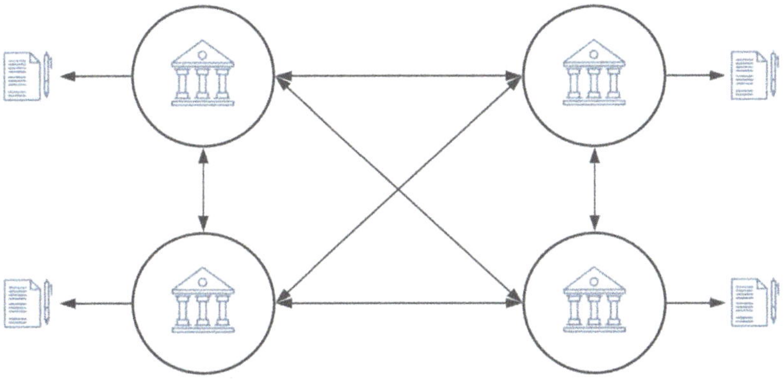

Figure 4.5. Distributed ledger.

various kinds of data transactions. Generally, this information is saved in the ledger merely when a consensus is arrived at by the groups concerned. Figure 4.5 shows a distributed ledger.

Whole files in the distributed ledger are then time-stamped and provided a distinctive cryptographic signature. Entire contributors in the distributed ledger could see all the entries in the query. The method presents a provable and ascertainable past of whole data saved in specific databases.

A distributed ledger displays all data and transactions to consumers and encourages lucidity [12]. It can reduce the time of the transaction to seconds and ensure billions of businesses are operating 24/7. The other technology benefits affect the part of a company's office that is not client-facing performance and mechanization. Distributed ledgers, for example, blockchain, could be very helpful for monetary transactions. They reduce workable inefficiency (which finally saves money). A high degree of safety is offered because of their decentralized nature, and the ledgers are unchanging.

4.4.1 Why is this one of the critical characteristics of blockchain?

Figure 4.6 shows why distributed ledger is one of the critical characteristics of blockchain.

- **No malicious changes** The distributed ledger responds efficiently to any suspicious activity or damage. Nobody could alter the ledger, and enhancements are rapid. So it is effortless to keep track of what is going on in the register at all of the nodes.
- **Right of confirmation:** The nodes operate as an authenticator of the ledger. If one customer needs to attach a novel module, others should check the transaction and provide the green signal. It offers genuine involvement for the customer.
- **No additional help:** anybody on the network could obtain distinctive assistance from the system. Everybody has to undergo the regular channels and afterwards attach their blocks. It is not like you have more power, so you obtain more offers.
- **Administration:** Each active node must keep a ledger and take part in the verification to implement blockchain characteristics.

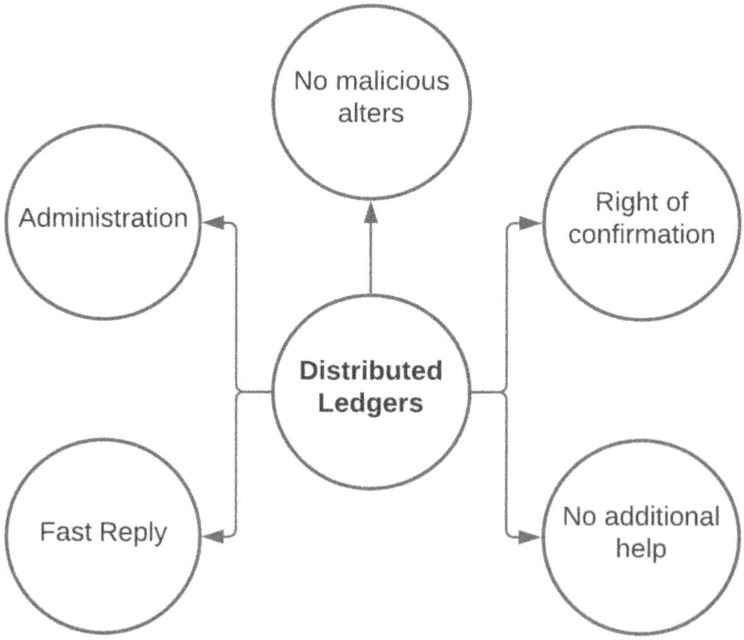

Figure 4.6. Reasons why the distributed ledger is one of the critical characteristics of blockchain.

- **Fast response:** As mentioned previously, detaching intermediates speeds up the computer response. Any changes to the ledger will register in seconds or minutes.

4.5 Consensus

Each blockchain prospers due to consensus mechanisms. The architecture is intelligently planned, and consensus mechanisms are at the centre of this structure. There is a consensus within each blockchain to create network decisions [13].

In simple terms, the consensus is the decision-creating procedure for a set of active nodes in a network. Therefore, the nodes can attain a consensus relatively speedily. When millions of nodes check a transaction, the consensus is utterly essential for a scheme to be executed without problems. It is analogous to a voting scheme where the majority wins, and the minority must assist it.

There is a consensus that the network is unreliable [14]. The nodes may not trust each other. However, they can trust the mechanisms that operate at its core. That is why every decision in the network is a successful scheme for blockchain. There are various consensus mechanisms for blockchains around the world. Each possesses a distinctive method of creating decisions and eliminates errors made in the past. Architecture generates a fair part of the web. But, there must be a consensus mechanism within each blockchain to pursue decentralization; otherwise, its core value is lost.

In simple terms, the consensus is an energetic method of arriving at conformity in a collection. Voting ensures that an agreement could reach an accord, while concurrently resolving the more than half rule in the absence of any thought about the emotions and welfare of the less than half, to the advantage of the collection as a whole. From a very ideological opinion, a collection of humans dispersed all over the globe could use consensus to generate a more matching and trustworthy society.

A technique of consensus choice-making is called a 'consensus method' [15]. Therefore, now that we have specified what consensus is, let us see what are the goals of the consensus method shown in figure 4.7.

- **Request agreement:** The consensus method is to come to as high a level of agreement as is feasible from the board.
- **Collaboration:** All contributors must aim to operate jointly to attain a decision that prioritises the greater interest of the collection.
- **Cooperation:** All contributors must not place their own interest above others and should operate as a team rather than as individuals.
- **Equality:** A collection attempting to reach a consensus must be as equitable as feasible. Based on this, each vote has equal weight. One person's vote cannot be more significant than another's vote.
- **Inclusivity:** Involve as many people as possible in the consensus procedure. It must not be susceptible to the irrational view that an individual's vote will not contain any weight in the long run.
- **Participation:** The consensus method is that everybody must actively take part in the all-inclusive procedure.

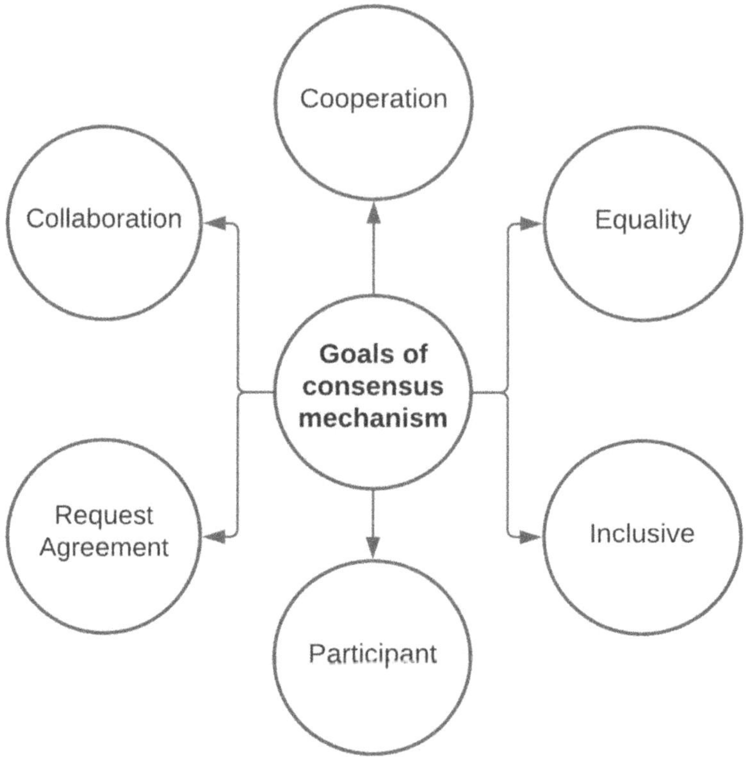

Figure 4.7. Goals of consensus mechanism.

4.6 Quick solution

Conventional banking methods are prolonged. Occasionally it could take days to operate a transaction such that all solutions have completed, causing a degree of distortion. Blockchain provides a quick solution compared to conventional banking methods. By this method a consumer could transmit cash comparatively fast thereby saving much time in the long run.

These blockchain characteristics make work uncomplicated for overseas employees and assist them in realizing why blockchain is necessary. Many leave their families to move to a different nation pursuit of a better life and work. But, transferring cash to their families at home requires time, and can be risky in times of need.

At present, blockchains are very quick, and can be utlized to transmit cash to loved ones. Another real benefit is the smart contracting method [16, 17]. This permits one to create immediate solutions to any contract. It is one of the significant advantages of blockchain to date [18]. Without third parties, a user could transmit cash with low fees.

Blockchain would also affect global trade in this method. Thus, consumers should utilize blockchain technology. However, there are a few cases where the network struggles to benefit more consumers; a quick fix is not feasible. Nevertheless, numerous consumers are enhancing the situation.

4.7 Summary

Blockchain technology is not just another hype that will be forgotten in time. With all the blockchain characteristics and appliances, we can safely say it is here to stay. All the critical features of blockchain have the effect of changing the web. Blockchain is integrated, with all kinds of novel technologies. Even though blockchain leads to many disputes, if we can harness the full advantages of blockchain, we could create an entirely lighter and smoother future. Additionally, blockchain will change the world. In this chapter, the various vital characteristics like immutability, decentralization, enhanced security, distributed ledgers, consensus and faster settlement are discussed. The next chapter discusses in detail the necessity of smart contract in blockchain technology.

References

[1] Arun M, Balamurali S, Rawal B S, Duan Q, Kumar R L and Balamurugan B 2020 Mutual authentication and authorized data access between fog and user-based on blockchain technology *IEEE INFOCOM 2020-IEEE Conf. on Computer Communications Workshops (INFOCOM WKSHPS)* (Piscataway, NJ: IEEE) pp 37–42

[2] Hofmann F, Wurster S, Ron E and Böhmecke-Schwafert M 2017 The immutability concept of blockchains and benefits of early standardization *2017 ITU Kaleidoscope: Challenges for a Data-Driven Society (ITU K)* (Piscataway, NJ: IEEE) pp 1–8

[3] Ye C, Li G, Cai H, Gu Y and Fukuda A 2018 Analysis of security in the blockchain: A case study in 51%-attack detecting *2018 5th Int. Conf. on Dependable Systems and Their Applications (DSA)* (Piscataway, NJ: IEEE) pp 15–24

[4] Puthal D, Malik N, Mohanty S P, Kougianos E and Yang C 2018 The blockchain is a decentralized security framework [future directions] *IEEE Consum. Electron. Mag.* **7** 18–21

[5] Yang Z, Yang K, Lei L, Zheng K and Leung V C 2018 Blockchain-based decentralized trust management in vehicular networks *IEEE Internet Things J.* **6** 1495–505

[6] Wang S, Zhang Y and Zhang Y 2018 A blockchain-based framework for data sharing with fine-grained access control in decentralized storage systems *IEEE Access* **6** 38437–50

[7] Halpin H and Piekarska M 2017 Introduction to security and privacy on the blockchain *2017 IEEE Eur. Symp. on Security and Privacy Workshops (EuroS&PW)* (Piscataway, NJ: IEEE) pp 1–3

[8] Fernández-Caramès T M and Fraga-Lamas P 2020 Towards post-quantum blockchain: A review on blockchain cryptography resistant to quantum computing attacks *IEEE Access* **8** 21091–116

[9] Ajao L A, Agajo J, Adedokun E A and Karngong L 2019 Crypto hash algorithm-based blockchain technology for managing decentralized ledger database in the oil and gas industry *Multidiscip. Sci. J.* **2** 300–25

[10] Bacon J, Michels J D, Millard C and Singh J 2018 Blockchain demystified: a technical and legal introduction to distributed and centralized ledgers *Rich. J.L. & Tech.* **25** 1

[11] Zia M F, Benbouzid M, Elbouchikhi E, Muyeen S M, Techato K and Guerrero J M 2020 Microgrid transactive energy: review, architectures, distributed ledger technologies, and market analysis *IEEE Access* **8** 19410–32

[12] Dai M, Zhang S, Wang H and Jin S 2018 A low storage room requirement framework for distributed ledger in the blockchain *IEEE Access* **6** 22970–5

[13] Zheng Z, Xie S, Dai H, Chen X and Wang H 2017 An overview of blockchain technology: Architecture, consensus, and future trends *2017 IEEE Int. Congress on Big Data (BigData Congress)* (Piscataway, NJ: IEEE) pp 557–64

[14] Mingxiao D, Xiaofeng M, Zhe Z, Xiangwei W and Qijun C 2017 A review on consensus algorithm of blockchain *2017 IEEE Int. Conf. on Systems, Man, and Cybernetics (SMC)* (Piscataway, NJ: IEEE) pp 2567–72

[15] Bach L M, Mihaljevic B and Zagar M 2018 Comparative analysis of blockchain consensus algorithms *2018 41st Int. Convention on Information and Communication Technology, Electronics and Microelectronics (MIPRO)* (Piscataway, NJ: IEEE) pp 1545–50

[16] Chen Y H, Chen S H and Lin I C 2018 Blockchain-based smart contract forbidding system *2018 IEEE Int. Conf. on Applied System Invention (ICASI)* (Piscataway, NJ: IEEE) pp 208–11

[17] Xiong W and Xiong L 2019 Smart contract-based data trading mode using blockchain and machine learning *IEEE Access* **7** 102331–44

[18] Niranjanamurthy M, Nithya B N and Jagannatha S 2019 Analysis of Blockchain technology: pros, cons and SWOT *Cluster Comput.* **22** 14743–57

IOP Publishing

Blockchain in the Industrial Internet of Things

Lakshmana Kumar Ramasamy and Seifedine Kadry

Chapter 5

Smart contract

Blockchain is an assuring technology that has attracted significant attention in both practical industrial appliances and educational research. The smart contract is a programmable transaction which could execute an innovative job, be automated and stored in the blockchain. A smart contract is a crucial element of blockchain that has developed blockchain as a technology that transcends the boundaries of electronic currencies and applies to appliances, for example, business process management, digital identification, supply chain, IoT, healthcare and more. This chapter presents and overview of the smart contract, its structure, how it is executed and its different aspects.

5.1 Overview of smart contract

Recently, advances in blockchain technology have allowed the customized logic system to maintain a decentralized method. This has stimulated the imagination and eased the development of smart contracts (further termed digital contracts, blockchain contracts, or artificial intelligence agreements) which were initially suggested by Nick Szabo in 1994 [1]. Smart contracts are self-employment agreements accompanied by the terms of the contract among the parties involved. Contracts are written in the structure of program codes which live on a shared, decentralized blockchain network. Smart contracts permit business dealings to take place among unscrupulous or anonymous individuals in the absence of a central authority [2].

The blockchain technology that represents bitcoin and different cryptocurrencies is termed blockchain 1.0. It has typical characteristics of decentralization, resistance interference, anonymity and auditing. But, signing sensitive contracts is impossible because of the restrictions of the bitcoin language (the bitcoin scripting language contains merely 256 instructions, of which 15 were recently disenabled, and 75 booked). Because of the restricted performance, bitcoin could be thought of as an example of smart contracts.

Recently, blockchain platforms, for instance, Ethereum [3] have held the concept of using user-defined programs in the blockchain, therefore generating smart contracts with the assistance of Turing-complete programming language. Ethereum smart contract codes are written in stack-based bytecode language and implemented in an Ethereum Virtual Machine (EVM). Numerous top-level languages like Solidity1 and Serpent2 can be utilized to write smart Ethereum contracts. The code for those languages may be compiled into EVM code that will work. Ethereum is currently the most famous platform for emerging smart agreements; that is why it is termed Blockchain 2.0 [4]. Besides Ethereum, other platforms can be used to create smart contracts; for example, Hyperledger Fabric [5], Corda [6], BigchainDB [7], and so on.

Contracts may incorporate some of the predefined regulations and perform the equivalent functions where the starting states are fulfilled [8]. Therefore, smart contracts could be used in a wide range of domains, including intelligent benefits and to enforce the independence of administration appliances (for example, digital asset management, for instance, ujomusic4, e-voting, and supply chain) [4].

In addition to the disclosure of smart contracts, the system faces more technical problems. A familiar example is where in June 2016, the 'Recursive calls attack' used a decentralized autonomous organization (DAO) to withdraw one-third of the DAO funds from a managed account (the affected Ether had an estimated value of $50M [9]). The Ethereum community had to use a heavy fork (which is a significant change in the agreement which makes formerly unauthorized blocks/transactions valid, or the converse) for Ethereum blockchain to recover the affected funds. However, this was contentious since it contravenes the code is law principle. Cheng *et al* [4] have recognized several analysis gaps in smart contract analysis, for example, lack of qualifications about performance and scalability problems, lack of research on dealing with offender actions, lack of implementation and running smart contracts outside of Ethereum, etc.

The principal purpose of this analysis is to provide a concept and architectural design of the smart contract, its future applications and styles, and so on. Figure 5.1 shows the architecture of a smart contract.

5.2 Operating methods of smart contracts

Smart contracts are a collection of visual-response practical logic and rules. Otherwise, they are decentralized, reliable distributed codes used in blockchain.

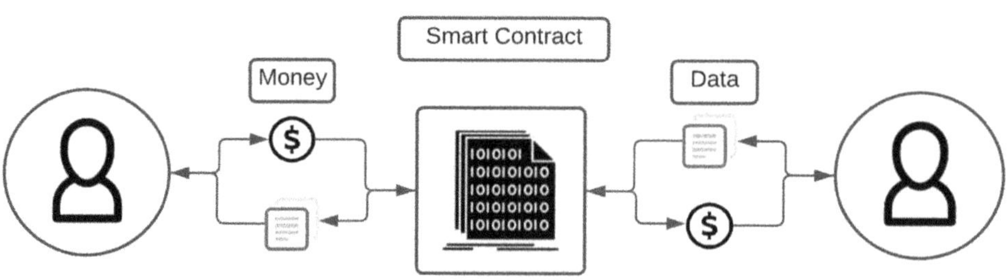

Figure 5.1. The architecture of smart contract.

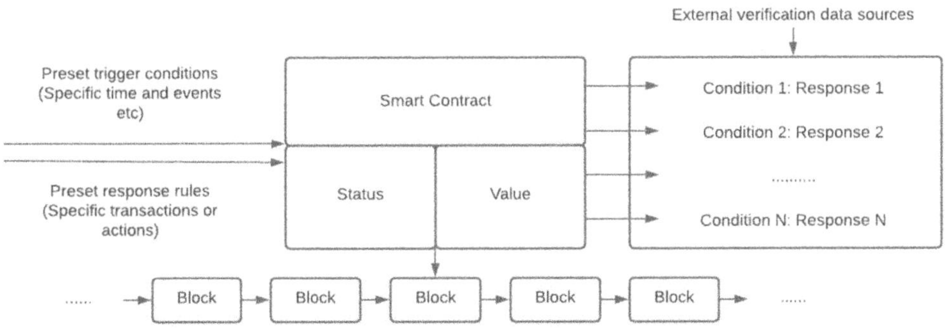

Figure 5.2. Operating methods of the smart contract.

The parties signing the contract must agree on the contract details, terms of breach of contract, accountability for breach of contract and external confirmation data sources (Oracles). The entire procedure is autonomous of any central agencies. The operating method of smart contracts is shown in figure 5.2.

Typically, smart contracts are linked to blockchain in the structure of program codes (e.g., a bitcoin transaction) after they are signed by all parties, and they are promoted by the P2P network and then registered on the blockchain after being confirmed by the nodes. The smart contract includes several predefined conditions and changes regulations, such as scenarios that trigger the contract implementation (for example at a specific time or a particular event), as well as replies to a particular situation. Blockchain watches actual-time condition and implements smart contracts and agreements after the exact stimulus status has been encountered [10].

5.3 Sites and programming languages

Smart contracts can be created and used on various blockchain sites. Numerous sites have numerous properties. This section introduces two common areas, that is to say, Hyperledger fabric and Ethereum.

- **Hyperledger fabric**: Hyperledger fabric is the function of a blockchain framework provided by the Linux Foundation. It is the site for distributed ledger resolutions, supported by the scalable structure which offers a large degree of secrecy, retraction, elasticity and measurement. It is planned to assist the pluggable processing of various elements and causes complication throughout the economic ecosystem. Hyperledger fabric container technology allows one to host smart contracts termed 'chain codes' that incorporate the appliance logic of the network. In addition, 'chain codes' is a channel which interrelates with blockchain and the source which creates transactions.
- **Ethereum**: Ethereum is the public blockchain site in which appliances work as planned in the absence of idle time, auditing, fraud, or third-party hindrance. The contract uses the Turing-complete virtual machine termed the EVM to work the bytecode (the bytecode is run through the solid as previously stated). Appliances in Ethereum work on its site-specific cryptographic token—Ether. Ether is utilized in two ways, one is buying and selingl goods and services

with a digital currency similar to bitcoin; the other is to provide recompense for contributor nodes for the calculations they perform [11].

As previously mentioned, the codes of Ethereum smart contracts are written in stack-based bytecode language and implemented in EVM. Many high-level languages (for example, Serpent, Solidity, and LLL) can utilize to write Ethereum smart contracts. The codes are assembled into executable EVM bytecodes. Hyperledger creates smart deals based on Fabric Java and Go. The spirit of the use of smart contracts in the Hyperledger fabric is to understand the three functions of the 'chain code' interface: start, call and query, that are contract order and placement, transaction inquiries and transaction processing, separately [12].

5.4 Features and the importance of smart contracts

Smart contracts have three features that are decentralization, self-sufficiency and autonomy. First, smart contracts are decentralized because they are not on a centralized server; they are shared over network nodes and are self-executing. Second, the smart contract may be independent in its capacity to marshal resources—i.e. increasing funds by offering services and expending them when required, for example, acquiring and executing energy or storage. Third, autonomy means that once they are started and implemented, contracts and start-up agents no longer need to be contacted [13].

Smart contracts have leading suggestions for blockchain. However, smart contracts are the operators of blockchain, which lays the substructure for the social and financial systems which could be programmed in the epoch of blockchain 2.0 and 3.0. The programmable characteristics enable the integration of nodes with complicated behaviours into a shared blockchain. They are making it feasible for the upcoming decentralized autonomous society (DAS), decentralized independent corporation (DIC) and decentralized autonomous organization (DAO).

5.5 Application views of smart contracts

Numerous application views can be used for the intelligent contract. A few of these applications are shown in figure 5.3, they are:

- **Financial transactions**: The smart contract is especially appropriate to companies, for example, online insurance, peer-to-peer (P2P) lending and equity crude funding. Conventional monetary trading should be interrelated through centralized firms, for example central settlement firms or interchanges,

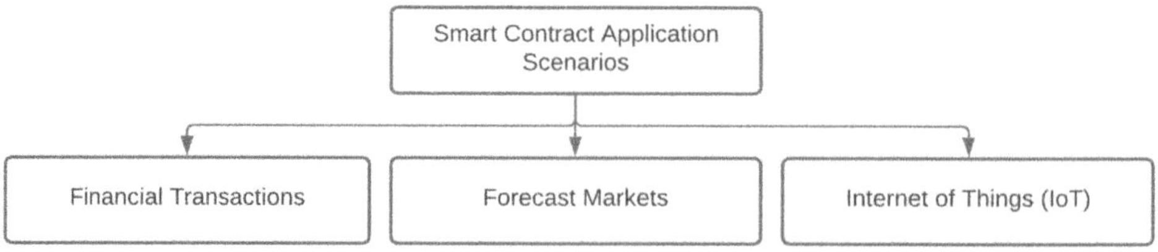

Figure 5.3. Smart contract application scenarios.

if the dynamic aspect of the smart deal could enormously decrease transaction prices and raise competence, therefore evading complex settlement and distribution [14].

- **Forecast markets:** Forecast markets have been verified to be capable of offering best upcoming predictions, many speculative mechanisms and direct hedging. Because of the consensus confirmation and consistency of the distribution, the smart contract can be utilized in the forecast markets. The two most common uses are Gnosis and Agur. Agur 8 manages to generate a surprisingly correct forecast tool with the assistance of blockchain. Gnosis offers frisky surroundings for contributors to attempt to trade in markets and gain GNO tokens like a prize for victorious forecasts. Additionally, the smart contract could be further utilized for gambling or voting [15].
- **Internet of Things (IoT):** The mixture of IoT and smart contracts does not merely facilitate the distribution of data among nodes, however, it further facilitates time-dependent mechanical workflows in a cryptanalytics realistic way. Slock.it suggested the initial independent lock that the user could unlock with a token. Lock owners determine merely two numbers—the price of the deposit and the price of the rent. The locked user scans it and locates what he or she has to reward using the guarantee. When receiving a deposit by lock, control is allowed. Users could unlock and close the lock whenever they need. When the user no longer requires the security, they obtain a refund of their deposit rent that will be transmitted to the lock owner. Apart from renting houses and apartments, the lock could further utilized for padlocks, bikes, cars, and so on. A smart contract could additionally use on the smart grid to buy and sell renewable power in the P2P market, therefore decreasing intermediary prices [16].

There are more application views, for example, intelligent transport, digital rights management of smart contracts, supply chain, cloud storage, social media sites and so on.

5.6 Challenges

Smart contracts have unique properties compared to conventional shared appliances. However, more smart contracting sites, for example, Ethereum, work on public networks that could be joined by arbitrary contributors. Conversely, because of the immutability of the blockchain, contracts cannot be exchanged as soon as they are utilized. Thus hackers could use this to attack. Additionally, privacy and security are dominant, like smart contracts commonly used to transmit digital wealth. Figure 5.4 shows a few of the challenges recently facing the smart contract:

- **Transaction-ordering dependency (TOD):** This happens while numerous necessary transactions are executing a similar agreement added to a block. We can see how miners could establish an arbitrary order among transactions, i.e. how the miner orders the transactions and the last conditions are

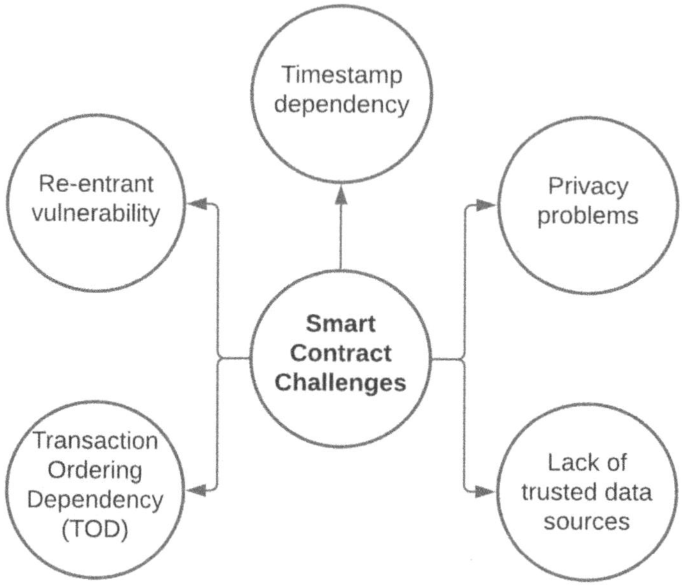

Figure 5.4. Smart contract challenges.

in agreement. Therefore, an enemy could favourably initiate an attack if those changes are not implemented in the correct order.

- **Re-entrant vulnerability:** This issue happens while a hacker uses a recursive call method to perform numerous removals when their stabilities are merely deducted one time. It could lead to unpredictable actions and finally even absorb the entire gas.
- **Timestamp dependency:** Normally, the timestamp is fixing to the recent time of the local organization of the miner. However, the miner could modify this value, when other miners agree on the block. The safety issue emerges while the timestamp is utilized as a trigger state for specific actions (for example, transmitting currency) since an attacker may use various block timestamps to handle the end of the contract. We term such an agreement a timestamp-reliant contract.
- **Privacy problems:** Because all transmission history is recorded in the blockchain and is known to any person, it is feasible to get the user's data by examining the transmission graph structures, which we term a deanonymization attack [17].
- **Lack of trusted data sources:** Smart contracts occasionally need data from outside sources. But, the authenticity of the data cannot be assured.

5.7 Current developments and upcoming trends of smart contracts

This section describes the current results and provides the forthcoming improvement trends of smart contracts.

5.7.1 Current developments

Analyzers have suggested numerous solutions to overcome the above challenges. Natoli *et al* [18] suggested, based on Ethereum-based functions (for example, SendIfReceived), to execute a sequence of transactions to evade TOD. Liu and others [19] presented a coded execution tool named Oyente to explore for possible safety bugs. They further suggested methods to improve Ethereum's functional linguistics to give contracts a smaller amount of risk. Targeting the absence of reliable information sources, Al-Breggi *et al* [20] suggested the Town Crier solution that presents reliable information for smart contracts from dedicated web servers. Liao *et al* [21] proposed a blockchain model named Hawk that does not explicitly record monetary transactions in the blockchain, therefore preserving users' transactional secrecy.

5.7.2 Upcoming trends: parallel blockchain and smart contracts

The fast growth of the internet recently and its deep connection with the physical world have altered the means of manufacture, human lifestyles and the administration manner of contemporary society. The growth trajectory of upcoming companies should take into account human and social components ranging from cyber-physical systems (CPSs) to cyber-physical social systems (CPSSs). Currently, the CPSS-based parallel community is starting to emerge, and its primary and vital features are feedback regulation and virtual-real communication.

Blockchain and its smart contracts are the fundamental infrastructures for understanding a CPSS-based parallel society. Its central participation is to present useful decentralized data formations, collaborative algorithms and computer replicas for shared social networks and shared AI, therefore establishing a formidable foundation for the upcoming parallel society. For instance, contemporary centralized social methods unavoidably have the characteristics of 'Merton systems' because of their considerable engineering complication and social complication, i.e. uncertainty, diversity and complexity (UDC). Therefore federal agencies and policymakers could make an illicit profit from their offer. Luckily, smart contracts could assist software in understanding limited social networks.

The fundamental idea is to eliminate centralized authorities, using changeable activity in the method of contract codes so that they are spontaneously implemented, making it harder to commit fraud. As a result, to a few areas, Merton's network could change into Newton's system, which could be extensively perceived, intensively managed, and precisely forecasted. The ACP technique (Artificial societies + computational tests + parallel execution) is the merely systematic and comprehensive analysis structure in the domain of similar management. The ACP technique could openly be linked to smart contracts, therefore realizing the parallel administration of socio-economic networks.

5.8 Advantages of smart contracts

Figure 5.5 shows some advantages of smart contracts.

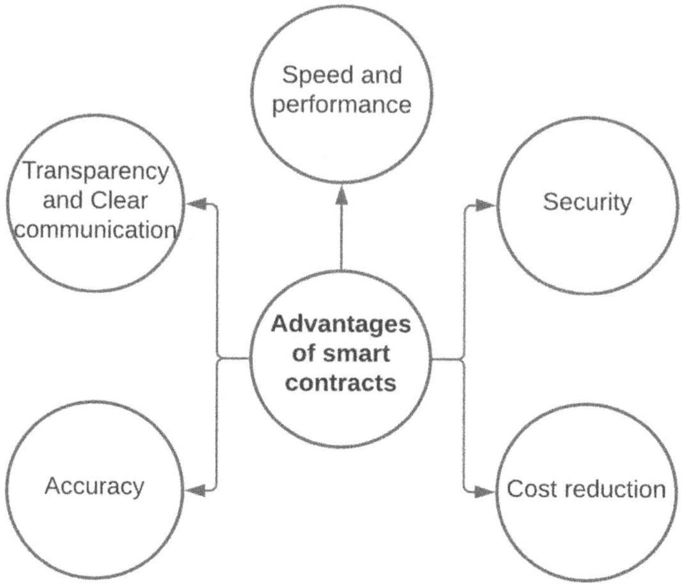

Figure 5.5. Advantages of smart contract.

5.8.1 Accuracy

Precision is one of the benefits that businesses can gain from implementing smart contracts. All data related to the agreement is disclosed in the conditional form, as described in the process of setting up the smart contract. For example, if user x rewards x unit y when commanding a particular service package, the beneficiary is instantly credited and the service package unlocked for user x because most contracts are subject to cash transfer.

Smart contracts could be synchronized with cryptocurrencies, for example, Etherium, light currency or bitcoin, a feature which additionally enhances the strength, accuracy and effectiveness of the whole network. The disclosure of fundamental rules and regulations contained in the smart agreement should be transparent and precise. It is a crucial need; however, transaction errors will emerge from any deletion. Thus, the automation exhibition on smart contracts avoids many of the problems established in conventional agreements.

5.8.2 Transparency and clear communication

The rules and regulations of the contract and terms are clear to various network players of a particular blockchain. Thus, once the agreement is accepted, exchanges can be effortlessly executed. Each transaction of both parties to the contract is monitored and controlled by other network nodes in the blockchain. Accordingly, transparency is encouraged, and abuse problems removed.

In the contemporary epoch, numerous cases have been announced in which the company has been accused of defrauding users and not giving them the value of their money. As mentioned earlier, every sale of a good or service by a company results in a contract that might not be lawfully executed. In different cases, one or more parties to the agreement might violate the rules and regulations of the contract. Depending on the sale, a company might charge a higher fee to the customer or reduce the agreed service package time without informing the customer. Implementing smart contracts sheds light on every detail of the contract. Different from the conventional arrangement, the company will have to utilize the lawful structure as a mediator. In the virtual world, all that is required is other nodes in the network that are responsible for guaranteeing each transaction. The contract is precise and valid.

5.8.3 Speed and performance

Smart contracts do not depend on human interference, and their execution is led and supervised by various nodes in the blockchain system. Thus, once the contract is induced, the written agreement runs automatically. It is frequently achieved by using influence events when script translating. For example, an induce event might be an act introduced by date, time, or party to a contract, for example, the transmission of some unit of cryptocurrency from the user's wallet to the organization. When an induce event occurs, the contract then begins to be executed alone. For example, for online subscription-based companies, the user subscription will automatically update upon receipt of a specific unit of cryptocurrency.

Different from conventional contracts that need low performance and a few kinds human confirmation, at the moment, the nodes in the blockchain system decide on the assurance of whether the accurate quantity has been paid, and whether the accurate subdivision, service and related features have been provided to the number. Therefore, there is no longer any dependence on the company's implemented network to decide contracts with users. The company has no authority ruling over transactions and legal agreements with partners. Each contract is selected at an individual company and each transaction is checked initially regardless of its beginning. It results in a quick, flexible and robust way to execute the contract.

5.8.4 Security

Smart contracts are one of the biggest security measurements. A smart contract developed via blockchain technology utilizes a decentralized system created by trusted parties. The parties in the network check each other to guarantee that each transaction is conveyed efficiently and that there is an equivalent global outlook on the condition of all transactions. Again, blockchain technology is developed via cryptography techniques. This technology uses public and private keys to encrypt information, read trades within each blockchain, and implement any transaction.

The security of smart technology is enhanced by the reality that before any node can make a transaction, the transaction should initially be verified by all the nodes over the blockchain network. The research discusses use of data encryption and, in particular, use of cryptographic techniques to considerably improve the security of communication and data transfer. Therefore, any contract executed in an encrypted method enhances the safety of the transaction and prevents malicious actions that might spread to changing the implementation order or carrying out incorrect transactions.

5.8.5 Cost reduction

High-level business managers are responsible for development of plans and methods to reduce costs within a company. The primary purpose of arranging a business is to create a profit; thus, all behaviours in a company should be structure in such a way as to encourage the attainment of corporate goals and increase the affluence of the stockholders. In the latest trend, which has witnessed a broader technological evolution, the success of businesses lies in their capability to keep track of existing technologies and to embrace commitment and techniques that raise employee productiveness and efficiency.

Implementing smart contracts via blockchain technology reduces the necessity for an intermediary such as lawful personnel. It helps to reduce overall corporate costs and increase the margin gain by a company. In the case of multinational companies who conduct a large number of contracts on a weekly or daily basis, enabling smart interactions with its business partners and users could considerably help reduce the different costs suffered by the conventional structure of contracts.

Contracts could also enhance a company's effectiveness that is an essential component of administrative success and raised productivity. It should be noted, however, that despite the safety, cost-cutting and effectiveness features related to smart contracts, they are not a magic bullet in every case and might therefore be subject to vulnerabilities. For example, the nature and implementation of the agreement depend on the input that is fundamentally the encoded version of the contract. Thus, if there are defects in the arrangement of smart contracts, such flaws might induce unfavourable results, and the nature of the outcome might be altered.

5.9 Summary

With the quick evolution of blockchain technologies, increasing use of smart contracts has become a significant analysis subject in both industry and education. The unchangeable and immutable properties of smart contracts enable transmission of intellectual property, stocks, money, and so on, in a straightforward, friendly method, without the intrusion of third parties. Therefore, smart contracts will broadly be utilized within social and financial networks in the future. This chapter presents the perspective of the smart contract, which includes its application scenarios, architecture and concept. This chapter further discusses the challenges facing the smart contract and outlines its upcoming trends. The questions of parallel blockchain and associated smart contract appliances are discussed in the forthcoming text. The next chapter discusses in detail the taxonomy of blockchain networks.

References

[1] Szabo N 2017 Smart contracts: building blocks for digital markets, 1996. URL: http://www.fon.hum.uva.nl/rob/Courses/InformationInSpeech/CDROM/Literature/LOTwinterschool2006/szabo.best.vwh.net/smart_contracts_2.html

[2] Zichichi M, Contu M, Ferretti S and D'Angelo G 2019 LikeStarter: a smart-contract based social DAO for crowdfunding *IEEE INFOCOM 2019-IEEE Conf. on Computer Communications Workshops (INFOCOM WKSHPS)* (Piscataway, NJ: IEEE) pp 313–8

[3] Yavuz E, Koç A K, Çabuk U C and Dalkılıç G 2018 Towards secure e-voting using the ethereum blockchain *2018 6th Int. Symp. on Digital Forensic and Security (ISDFS)* (Piscataway, NJ: IEEE) pp 1–7

[4] Cheng J C, Lee N Y, Chi C and Chen Y H 2018 Blockchain and smart contract for a digital certificate *2018 IEEE Int. Conf. on Applied System Invention (ICASI)* (Piscataway, NJ: IEEE) pp 1046–51

[5] Sukhwani H, Wang N, Trivedi K S and Rindos A 2018 Performance modelling of hyper ledger fabric (permissioned blockchain network) *2018 IEEE 17th Int. Symp. on Network Computing and Applications (NCA)* (Piscataway, NJ: IEEE) pp 1–8

[6] Singh G, Chelini L, Corda S, Awan A J, Stuijk S, Jordans R, Corporaal H and Boonstra A J 2018 A review of near-memory computing architectures: opportunities and challenges *2018 21st Euromicro Conf. on Digital System Design (DSD)* (Piscataway, NJ: IEEE) pp 608–17

[7] Orjuela K G, Gaona-García P A and Marin C E 2020 Towards an agriculture solution for product supply chain using blockchain: case study Agro-chain with BigchainDB *Acta Agric. Scand. B: Soil Plant Sci.* **71** 1–6

[8] Chalaemwongwan N and Kurutach W 2018 Notice of violation of IEEE publication principles: state of the art and challenges facing consensus protocols on the blockchain *2018 Int. Conf. on Information Networking (ICOIN)* (Los Alamitos, CA: IEEE Computer Society) pp 957–62

[9] Mehar M I, Shier C L, Giambattista A, Gong E, Fletcher G, Sanayhie R, Kim H M and Laskowski M 2019 Understanding a revolutionary and flawed grand experiment in the blockchain: the DAO attack *J. Cases Inf. Technol.* **21** 19–32

[10] Delmolino K, Arnett M, Kosba A, Miller A and Shi E 2016 Step by step towards creating a safe, smart contract: lessons and insights from a cryptocurrency lab *Int. Conf. on Financial Cryptography and Data Security* (Berlin: Springer) pp 79–94

[11] Pierro G A and Rocha H 2019 The influence factors on ethereum transaction fees *2019 IEEE/ACM 2nd Int. Workshop on Emerging Trends in Software Engineering for Blockchain (WETSEB)* (Piscataway, NJ: IEEE) pp 24–31

[12] Luu L, Chu D H, Olickel H, Saxena P and Hobor A 2016 Making smart contracts smarter *In Proc. of the 2016 ACM SIGSAC Conf. on Computer and Communications Security* pp 254–69

[13] Portmann E 2018 Rezension, blockchain: blueprint for a new economy *HMD Praxis der Wirtschaftsinformati* **55** 1362–4

[14] Zeng X, Hao N, Zheng J and Xu X 2019 A consortium blockchain paradigm on the hyper ledger-based peer-to-peer lending system *China Commun.* **16** 38–50

[15] Poongodi M, Sharma A, Vijayakumar V, Bhardwaj V, Sharma A P, Iqbal R and Kumar R 2020 Prediction of the price of Ethereum blockchain cryptocurrency in an industrial finance system *Comput. Electr. Eng.* **81** 106527

[16] Christidis K and Devetsikiotis M 2016 Blockchains and smart contracts for the internet of things *IEEE Access* **4** 2292–303

[17] Béres F, Seres I A, Benczúr A A and Quintyne-Collins M 2020 Blockchain is watching you: profiling and deanonymizing ethereum users arXiv preprint arXiv:2005.14051

[18] Li J, Gu C, Wei F and Chen X 2019 A survey on blockchain anomaly detection using data mining techniques *Int. Conf. on Blockchain and Trustworthy Systems* (Singapore: Springer) pp 491–504

[19] Liu C, Liu H, Cao Z, Chen Z, Chen B and Roscoe B 2018 Regard: finding reentrancy bugs in smart contracts *2018 IEEE/ACM 40th Int. Conf. on Software Engineering: Companion (ICSE-Companion)* (Piscataway, NJ: IEEE) pp 65–8
[20] Al-Breiki H, Rehman M H, Salah K and Svetinovic D 2020 Trustworthy blockchain oracles: review, comparison, and open research challenges *IEEE Access* **8** 85675–85
[21] Liao D Y and Wang X 2017 Design of a blockchain-based lottery system for smart cities applications *2017 IEEE 3rd Int. Conf. on Collaboration and Internet Computing (CIC)* (Piscataway, NJ: IEEE) pp 275–82

IOP Publishing

Blockchain in the Industrial Internet of Things

Lakshmana Kumar Ramasamy and Seifedine Kadry

Chapter 6

Taxonomy of blockchain systems

Blockchain is the disseminated, freely available, unchanging, public digital ledger that is disseminated between network peers. This chapter provides a comprehensive taxonomic debate of fundamental notions for blockchain technologies appliances. Figure 6.1 demonstrates the taxonomy of blockchain for artificial intelligence (AI), classified concerning decentralization of AI algorithms and functions, blockchain type and formation, and the basic consensus algorithms used for decentralized dispersed transaction verifications over vital networks.

6.1 Decentralized applications of AI

AI appliances run separately to make knowledgeable decisions by implementing various arrangements, exploration, enhancement, learning, and knowledge detection and knowledge administration plans. Hoewever, the decentralization of the AI function is a complicated and formidable task. Figure 6.2 shows decentralized applications of AI.

(1) **Automated computing**
One of the primary purposes of AI appliances is to permit partially (or fully) independent functions of many smart agents (for example, lightweight software) by perceiving their module surroundings, protecting their internal state, and performing specific activities appropriately [1]. Present computer systems must manage huge diversity across all verticals, containing application interfaces, data storage systems, data processing systems, devices and data sources to run automatically. Implementing multi-layer strategies across every vertical not only helps to overcome diversity, but also assists in establishing intra-layer and inter-layer functionality over systems [2]. Blockchain architecture could act as an essential feature in guaranteeing functional localization, permanently monitoring communications among systems, devices, appliances, data and users, leading to the evolution of fully decentralized independent systems.

doi:10.1088/978-0-7503-3663-5ch6

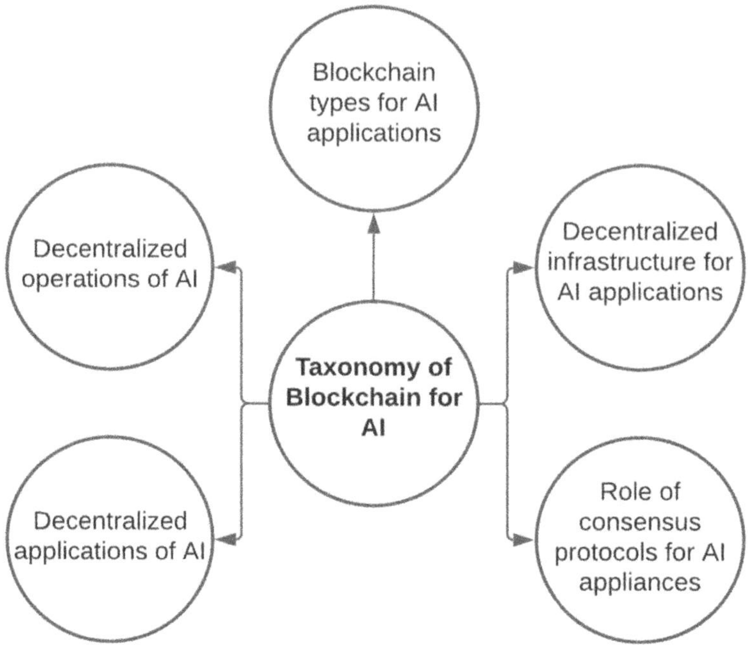

Figure 6.1. Taxonomy of blockchain for AI.

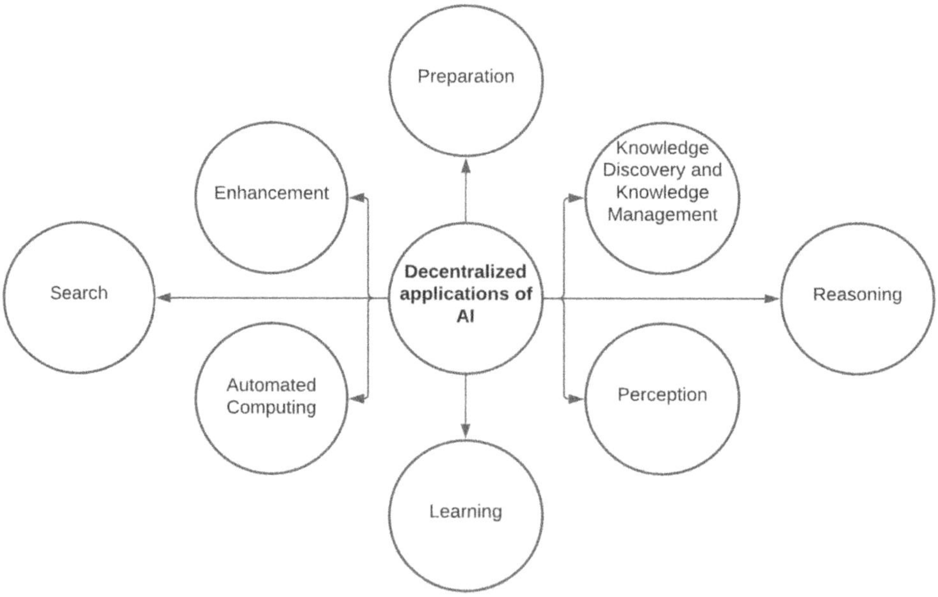

Figure 6.2. Decentralized applications of AI.

(2) **Enhancement**
One of the critical features of AI-allowed appliances and networks is to find a better feasible solution from total viable solutions [3]. Present AI appliance and networks are widespread,with omnipresent surroundings (for example, edge computing systems), resource-constrained surroundings (for instance, mobile systems/devices), physical features of area controlled systems (for example, PAN, WLAN and so on), and centrally mostly parallel and shared computer systems (for example, cloud computing systems) [4]. Enhancement plans operate in controlled or uncontrolled contexts, both system-level and application-level goals [5]. These plans help select the more appropriate data sources for a wide range of contexts, for example, the better contestant margin for data and appliances operating, or the better resolutions for allowing resource-proficient data administration in cloud servers or huge-scale shared computer surroundings. Current enhancement plans implemented with centralized control in mind are for system-level/appliances-level enhancement goals, resulting in inappropriate data execution and less computer/appliances effectiveness [6]. Implementing decentralized enhancement plans based on blockchain opens a novel window of analysis and growth opportunities. Decentralized enhancement leads to raising system effectiveness by executing the more appropriate data. Decentralized enhancement is advantageous when numerous plans operate at the same time, with various enhancement goals in appliances and networks.

(3) **Preparation**
AI applications implement preparation plans to cooperate with different applications and to resolve difficult issues in novel surroundings. Preparation plans contribute to the operational effectiveness and regression of AI applications by adopting the present input position and implementing the various rule and logic-based mechanisms to achieve pre-defined objectives [7]. Current centralized preparation is difficult and time consuming work; thus, blockchain-based decentralized AI preparation plans are required to provide many robust methods accompanied by constant monitoring and resource history. Blockchain can further help categorize essential and unchanging strategies for plan applications and mission-critical applications.

(4) **Knowledge discovery and knowledge management**
Present AI applications manage the massive quantity of data streams and need to assist in extensive centralized data executing systems. Centralized knowledge detection and knowledge administration application would be useful to present measured and computer-level ability; however, applications implement custom-built knowledge formats for particular sets of systems, devices, applications and customers [8]. The decentralization of knowledge detection procedures and decentralized knowledge administration are planned to present custom-built knowledge methods regarding

requirements of all shareholders in the system. Additionally, blockchain technologies ease the safe and detectable movement of knowledge between various stakeholders in AI systems and applications.

(5) **Perception**
Intelligent bots and agents in AI systems and applications repeatedly gather, explain, choose, and arrange data from their surroundings based on centralized perception techniques, resulting in the monolithic data group [9]. Decentralized perception plans help to gather data from various angles. Detection of blockchain-based diffusion perception pathways, enables safe transmission of saved data and unchanged data storage. Decentralized perception plans are helpful since systems and applications do not require continuing gathering of data streams for positive and high-quality impressions. Regarding the permanence of blockchain, merely the trace of positive feelings must produce confidence in blockchain.

(6) **Learning**
At the centre of AI algorithms will be learning algorithms to implement automation and knowledge detection procedures. Learning modalities differ based on exchange, reinforcement, and ensemble, semi-supervised, non-supervised, supervised, and in-depth learning techniques. These learning techniques often resolve various machine learning issues for machine mining, from categorization to clustering and regression research. Traditional learning techniques are trained based on a centralized framework to attain worldwide intelligence. In present AI applications [10, 11] decentralized learning techniques may assist in achieving more shared and independent learning applications, which helpfully integrate local intelligence over all verticals. Additionally, blockchain implements a stable and safer version of learning techniques by keeping historical features and data source. However, regarding the permanence of smart contracts, learning techniques need to be trained and tested before they can be used in the blockchain.

(7) **Search**
AI applications must work over huge and scattered search intervals (i.e., large datasets or diversified high-dimensional data streams); thus, proficient search plans become the substance of AI technologies. Search methods planned with a variety of elements in mind, for example, completeness, complication (for example, space and time) and optimization. These plans typically work on non-linear data structures, for instance, trees and maps, and the algorithms begin their enlargement anew, step-by-step expanding until they find the desired variable or travel across all search intervals. In general, search plans are executed based on huge-scale centralized and shared infrastructure to increase running productivity [12]. But, their execution based on decentralized infrastructure requires careful research. Rather than using critical search plans, it is expected that blockchain and decentralized infrastructure would be utilized to permanently and securely

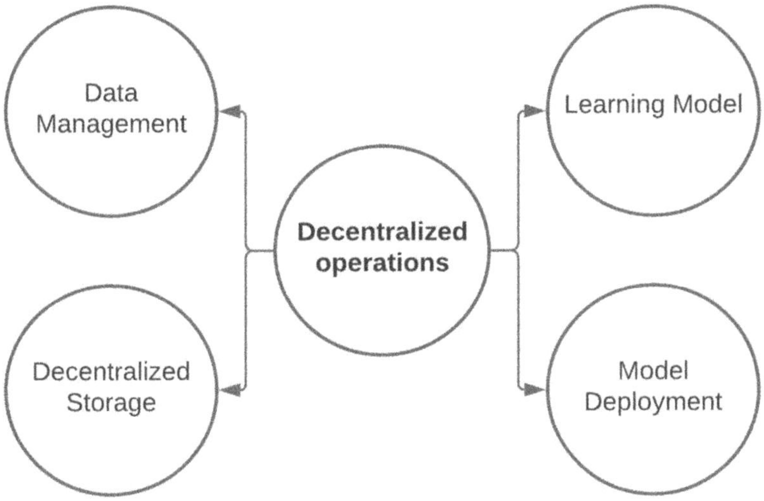

Figure 6.3. Decentralized operations.

save successful search clues and travel routes that may present optimum search resolutions for the same functions in the future.

(8) **Reasoning**
Logic programming is a crucial element of AI appliances which permits one to create exciting or exceptionally rational rules to achieve results. Rationalization of AI appliances leads to universal conduct common to all appliance elements [13]. Blockchain-based shared reasonable plans are designed to enhance highly customized logical programs that are most useful throughout feedback, learning, and sample arrangement to deal with this problem. Additionally, with smart contract-based decentralized, shared reasoning on blockchain ensures the accessibility of memorable rational procedures that will enable the implementation of the same logical master plans in the future.

6.2 Decentralized operations

AI appliances usually manage a considerable quantity of data for best and more adaptable decision making. However, a centralized database based on clouds, data centres, and clusters will begin to be a crucial obstacle to the progress of very safe and privacy-protected AI appliances. Additionally, model progress and arrangement learning are problematic in a few circumstances. Figure 6.3 shows decentralized operations.

(1) **Decentralized storage**
Centralized data storage is most easily compromised concerning security and privacy when it contains sensitive and personal data concerning financial information, health records, activities, locations, and users.

Additionally, a huge-scale data group reveals problems associated with the scaling and efficiency of a centralized framework, where AI appliances are required to operate, modify and save large amounts of data. Blockchain-based decentralized storage framework facilitates cryptographically safe data storage over contributing networks.

In these schemes, each node can have a customer-centric node. Respectable customers could mine and utilize their own data whenever required. Swarming and sharding are the main resolutions for decentralized storage. Sharding is the operation of generating reasoning division of databases through which a unique key is allocated to each division and utilized to access it. The shards are integrated, and their stored storage is assisted in bulk format by a node in the network. This enables parallel AI data access from numerous nodes on the network to decrease delays of total AI appliances. Additionally, the reliability and scalability of storage increases due to the terrestrial shared diversified storage schemes.

(2) **Data management**

Additionally, proficient decentralized storage, AI appliances must handle the data gathered from dependable data sources with very pertinent, precise and absolute databases. Conventionally, AI appliances operate centralized data administration programs running on all nodes in the base network [14]. These plans contain intelligent data management programs, cache data management, environmental awareness data storage and routing on necessary networks, data segmentation and data filtering. Regarding decentralized storage networks and uninterrupted barriers in blockchain, centralized data administration begins to be more ineffective. Since it would not merely grow a data copy if small modifications are made to the data, it would further require repetitive databases.

For large datasets, this big data move can lead to rapid bandwidth load and increased backhaul network traffic; thus, decentralized data administration is necessary for blockchain-based AI appliances. Decentralized data administration plans should be used at the node stages within the network regarding the spatial and temporal properties of the data. Additionally, decentralized data administration programs may place metadata on the blockchain to keep data and resources safe. Real data can save on conventional considerable storage schemes, for example, clusters and cloud data centres. For customer-centric small datasets, accurate data and metadata are kept in the blockchain and handled based on token-based privileges for nodes that store various pieces in the data network or participate in swarms.

(3) **Learning model**

Developmental learning in AI appliances plays a vital role in understanding the context from existing data and in creating instructed decisions using novel data. Typically, learning methods are trained and tested before real arrangement in existing systems. Centralized training is expensive since methods have to learn from full databases and are subject to excessive adaptation (for example, producing releases using remember data), but, contemporary AI appliances have to deal with continually developing data streams. The decentralization of the learning method evolution procedure is a viable method for creating resource-proficient learning methods for customer appliances [10].

Decentralized learning methods obtain input databases from appropriate clusters and fragments and create much-customized learning methods for each customer. Additionally, blockchain technologies can be used to maintain a data source and to create a consistent history of learning methods over time. Also, decentralized learning methods can begin to be a code for the same customers in the blockchain network, where customers are not require to train a novel approach. Once proficiently trained, learning methods can be tested based on several smart contracts in decentralized appliances.

(4) **Model deployment**

The actual effectiveness of the training method is evaluated after use in manufacture surroundings. However, the model arrangement is a continuous and operational procedure through which developers must continuously enhance techniques and adjust dependencies (for example, create a particular result by neglecting the remaining effects) to construct the most efficient and informative products. Model arrangement in centralized networks is an uncomplicated operational procedure, but, it is not easy when used in decentralized systems [15]. Smart contract-based model arrangement solves these difficulties by continually storing changes and keeping a stable version of various methods. Additionally, model distributing between various AI appliances is safe and reliable because developers can trace the source and all tracks of any particular version of a technique.

6.3 Blockchain types

Blockchain technologies classified as licensed (that is, only authorized customers can use blockchain appliances on private, federal, or cloud-based systems) or unlicensed (that is, publicly usable to all customers through the internet) systems. Figure 6.4 shows blockchain types.

(1) **Public**

A public blockchain is so-called unauthorized systems where customers can utilize, change, and download the blockchain code on their own computers. Additionally, public blockchain is attainable and open to reading and writing activities by all contributors in the network. Because of their accessibility, customer identifications and transferable privacy data on this blockchain are handled based on unknown and nickname data on the network. Moreover, this standard blockchain utilizes complicated systems for safety and consensus. The property and data exchanges in this

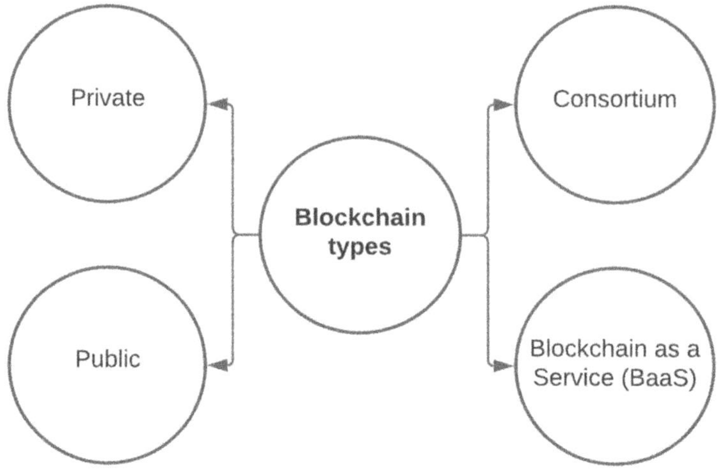

Figure 6.4. Blockchain types.

blockchain use their own tokens (that is, value pointers or cryptocurrencies) for each standard blockchain.

A public blockchain is broadly accepted because of its enormous distribution and accessibility, however, contributors (that is, validators/customers) are not always identified in this blockchain. Therefore, this blockchain is continually subject to malevolent safety attacks, which could lead to enormous value and information theft. Public blockchain requires the consensus of a minimum of 51% of validators and utilizes complicated mathematics to break safety codes, resulting in colossal power expenditure. Hackers are more likely to attack if they manage 51% of validators on the system. The transaction acceptance time on a public blockchain is comparatively lengthy in contrast to the consortium and private blockchain. Generally, a transaction in the public blockchain is agreed in 10 minutes or increased. However, this acceptance time relies on the number of contributors to the system.

(2) **Private**

A company handles a private blockchain. Unlike a public blockchain, a private blockchain is designed to be licensed organizations where customers/contributors already have the authority to write/read activities on computers and are always recognized on the network [16]. A private blockchain is relatively fast because of the already identified netizens and recognized identifications of the contributors in the structure; thus, a smaller number of complicated statistical functions are required to verify transactions in the network. Additionally, a private blockchain can transmit any local data, values and benefits to the system. The acceptance of commerce and property transactions is handled based on voting or multiple consensus mechanisms that use less power and make quicker transactions. In a private blockchain, the transaction acceptance time is often less than a second.

(3) **Consortium**
A team runs a consortium blockchain (further known as federal blockchain). Groups generally form using the mutual interests of the contributing companies [17]. Various factions, for example, government agencies, banks and private blockchain organizations, provide multiple kinds of federal blockchain. As a private blockchain, the federal blockchain runs as a licensed system, but, some contributors can write and read on the blockchain. Generally, all contributors in the network can read the information in the blockchain, however, only trusted and authorized customers can write the story in the blockchain. A federal blockchain is speedier than standard blockchain since contributors will always be recognized in advance accompanied by known identifications. Additionally, this blockchain uses low power due to consensus protocols using voting or multiple consensuses. A typical transaction in the federal blockchain is recognized within a second.

(4) **Blockchain-as-a-service (BaaS)**
Cloud service providers are concentrating on blockchain technologies because of the massive adoption and approval of giant corporations and governments. Leading cloud sellers such as IBM, Amazon and Microsoft use their environment to create and test blockchain services for their clients [18]. The rise of BaaS is considered an advantage over consortium and private blockchain organizations whose main focal point is on value inclusion via storage and computer infrastructure, testing and deployment regardless of infrastructure and application development. BaaS's functionality leads to novel cross-industry public–private partnerships, and to new business opportunities and business–customer communication methods. Developers are authorized to provide one-click BaaS services to write smart contracts. BaaS's combination accompanied by AI services opens up novel global opportunities for appliance developers as leading cloud sellers offer in advance a wide range of cloud services for AI appliances.

6.4 Decentralized infrastructure for AI applications

Blockchain structures have conventionally been planned as a linear framework using hashing plans and linked list data structures. However, the linear framework is developed using graph theory and sequence data methods to meet the necessities of actual-time appliances and to manage extensive data. Figure 6.5 shows a decentralized infrastructure.

Figure 6.5. Decentralized infrastructure.

(1) **Linear infrastructure**
Single chain-based blockchain structures progress in a straight line through which novel blocks are added at the end of the chain. Initially, decentralized networks operated on available chains, but, these networks had many problems. Single chains are measured gradually and illustrate the actual-time effectiveness of decentralized appliances. Additionally, each business environment requires distinct chains, so the exchange of value, data and property in numerous chains is unfeasible. Single-chain blockchain for AI appliances can be utilized for single-task AI appliances that work independently in exploration, enhancement and learning or similar contexts. Rather than running AI appliances based on smart contracts, single-chain blockchain is most effective when only effective histories are continuously saved. For instance, the in-depth learning method for diagnosing liver cancer in radiological appliances provides correct outcomes. Another example of the successful explores the path of remote industrial robots. Because AI appliances generally run in an open environment, placing all AI appliance elements in the blockchain is not a viable option.

(2) **Non-linear infrastructure**
Non-linear blockchain structures executed in the formation of multi-chain structures, in which blockchain topography is utilized in the construction of parallel chains, leading side chains and parent–child chains [19]. Multi-chain forms are not only measurable for actual-time effectiveness, but also assist different business plots and cross-chain value shifts. In multi-chain structures, one or more chains act as the main chain(s) with data regarding the other chains. The remaining chains act as parallel, child or side chains. The side and child chains generally function identically, however, in child chains, the business scenes are firmly attached to the parent chains. However, the side chains could run entirely autonomously from the leading chains. Parallel chains work autonomously from other chains. Value shift among various chains is done based on the 'pegging' method, where the two-way peg procedure is implemented at a constant transfer rate for bilateral value transfer between chains.

The tokens or local currencies denote the swap value in the blockchain. Blockchain for AI appliances permits numerous decentralized actions to perform interconnected or autonomous AI jobs. Additionally, scalable characteristics allow AI appliances to run in parallel during evolution and arrangement stages. In the manufacturing surroundings, AI elements are used in the parent/main chain, while training and testing appliances used inside chains or testnets. Major model non-structures are seeing increased application, for example, augmentation and adaptation learning methods that require to be continuously updated to correct their effectiveness through learning methods. In that event, the learning methods are created in the side chains and used in the leading chains.

6.5 Role of consensus protocols for AI appliances

This subsection provides consensus protocols and how they affect the performance of AI appliances in the blockchain. Figure 6.6 shows the role of consensus protocols for AI appliances.

(1) **Proof of work (PoW)**

PoW is a pioneering consensus protocol suggested by Satoshi Nagamoto, unknown founder of cryptocurrency and decentralized distributed ledger technologies. Famous public blockchain networks, for example, Ethereum and Bitcoin use a consensus protocol to verify transactions after contribution in a minimum 51% of nodes on the base network. Because the verification nodes work anonymously and on a considerable scale, these nodes must be mined by resolving a complicated and random statistical issue and breaking the hash code to read transactions in the blockchain. Successful nodes send the resolution over a peer-to-peer network to acquire rewards.

Novel transactions and data will be continuously attached to the blockchain when 51% of the nodes in the network profitably resolve the math issue. Even though PoW has proven to be the most acceptable consensus protocol, it uses a lot of power on massive systems and introduces delays in transaction acceptances. AI appliances have a higher frequency of writing processes since smart algorithms for smart results are continually updating the final structures. Thus, the effectiveness of PoW protocols in actual-time AI appliances is complex. Additionally, 51% of attackers on the base network may compromise the safety of AI appliances.

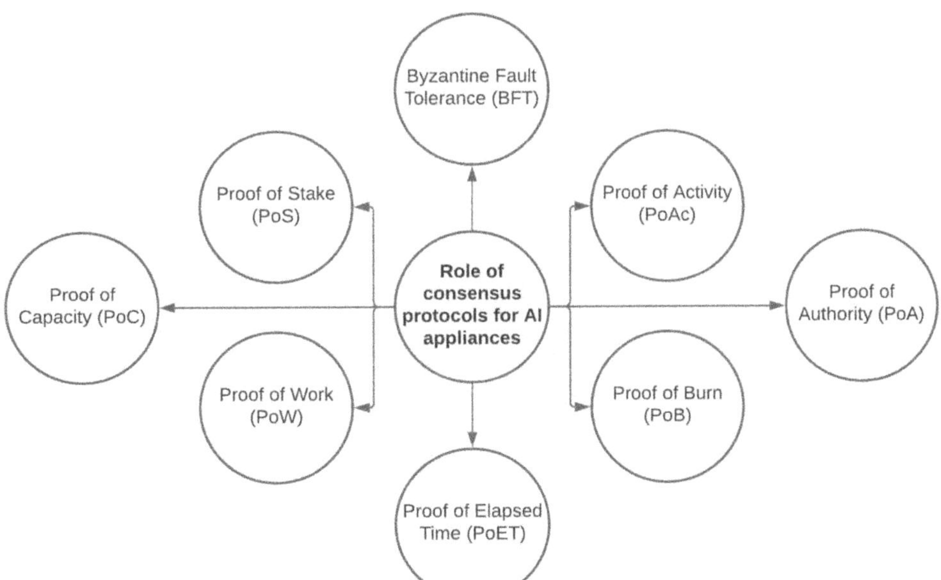

Figure 6.6. Role of consensus protocols for AI appliances.

(2) **Proof of stake (PoS)**
PoS-based consensus protocols resolve PoW's huge power expenditure problem. PoS protocols are run by explaining large stakeholders in the blockchain network and enabling them to generate novel modules. These protocols choose validators using various criteria (that is, long-term currency holders or huge frequency exchange validators, delegated validators and random validators). PoS has proven to be more powerful compared to PoW, and it incidentally resolves the safety issue by stopping unknown validators and only enabling validators who possess blockchain's own currency. But, if the transactions are not verified, the checkers have nothing to mislay in the blockchain; thus, delays can occur when generating novel modules. PoS should be helpful for late AI appliances; however, these protocols do not apply to AI appliances that manage flowing data, discover changes, and make actual-time knowledgeable decisions.

(3) **Byzantine fault tolerance (BFT)**
The BFT is the more significant number of a voting system that rejects verifications from malevolent nodes in the blockchain network [20]. Hostile nodes are a fraction of the blockchain, but can hold malicious intent code, which could directly/indirectly lead to wrong verifications and corrupt the data saved in the blockchain. Because entire nodes are a fraction of blockchain networks, detecting these malicious nodes is a challenge to the BFT protocol. Although hard to implement, BFT algorithms have long been utilized in critical systems, for example, nuclear systems, huge-scale sensor systems, and aircraft engines. Various types of BFT protocols are used. Easy BFT manages error tolerance for the reason that there are a minimum of two-thirds error-free nodes in the network. More BFT algorithms deal with fault tolerance by enabling digital signatures and limiting communication among peer nodes in the system. The BFT-based consensus is easy for AI appliances regarding the successful execution of BFT algorithms on complex computers.

(4) **Proof of activity (PoAc)**
The PoAc protocol is a mixture of PoS and PoW. This protocol primarily runs on blank blockchain based on the PoW method and resolves 51% of the attack issue. Firstly, the PoAc protocol resolves complicated math issues and verifiers begin to receive awards that raise their role in the blockchain. The protocol then implements the PoS algorithm for validators with fair shares on the blockchain. PoAc has been verified to be useful in safety, network connections and storage. Thus, it is helpful for AI appliances that require low data accessibility and high security.

(5) **Proof of burn (PoB)**
The PoB protocol enables validators to expend their currency merely by transmitting it to an inaccessible, verifiable, public, and invalid address. Customers will be permitted to generate novel blocks and will be rewarded

as soon as they burn the coins. PoB has the advantage of enabling customers to spend firstly cause their shares on blockchain and begin to be authorized validators. It further resolves PoW's power usage issue. Additionally, the coin burning technique decreases the number of coins in the blockchain; thus, the value of the currency increases step-by-step. The advantage of burning cash is that it balances the number of coins in the blockchain, spends unsold coins and pays the transaction fees on the network. AI appliances may use PoB protocols if they wish to motivate customers to keep the value of immediate results. For instance, an application that supports a particular stage of precision could find a specific number of clusters or a minimal number of objects, learning models and search trees.

(6) **Proof of elapsed time (PoET)**
Rather than involving all customers in the verification procedure, PoET protocols discover a leader who could generate novel volumes in the chain. A PoET protocol is implemented by attaching a random timer to each node on the network and selecting the minimal expiration node leader [21]. LeaderNote generates novel blocks and sends its signature to the entire system. A PoET protocol regularly implements the random leader chosen procedure, and novel leaders can be found all the time. It further helps to detect malevolent customers if similar nodes are selected as leaders or the minimal timer value is often allocated to particular nodes. PoET resolves the power expenditure issue of PoET; however, because of random time allocation, AI appliances may change slowly because the system will have delays up to when it expires. PoET can be helpful when utilized with a delay-tolerant application.

(7) **Proof of capacity (PoC)**
Conventional PoW methods are mathematically meticulous since they require random nonce values to open blocks. The POC protocol, further recognized by proof of space, acts as a substitute protocol for locating hard drive space at the ends of a blockchain network. Instead of random creations, it saves real possible nonce values on the hard drive and discovers incompatible nonce–hash pairs to open blocks. Based on PoC, nodes with huge disk space obtain an extra stake accompanied by a higher probability.

(8) **Proof of authority (PoA)**
PoA resolves the issue of colossal power expenditure of PoW. PoA protocols further resolve the issue of PoS dependence, so validators should hold a cash stake in the blockchain. A PoA protocol operates by giving official control over particular nodes that create a consensus using a significant number of votes to generate novel blocks in the network [22]. PoA has proven to be a power proficient and low delay consensus protocol; however, it is best suited for private networks to provide verification power to legal partners. Thus, blockchain operators should regard the legitimate identification of validators, clearly describing the ability criteria to play a

role as a validator and global ability criteria for acting as validators for all shareholders. Despite their power effectiveness and price efficacy, safety dangers to the PoA are always huge because of safety attacks on validators that can begin to be a source of attack over the network. But, PoA can be utilized as an alternative consensus protocol for AI appliances used on consortium or private networks because complete verifications are recognized throughout the system.

(9) **Proof of importance (PoI)**
The PoI algorithms are identical to their PoA number in that the verification nodes are rated based on the frequency of successful verifications. Validators with higher frequency gain more prominence in the blockchain and preserve their authorized transactions or other validators in the network since PoI puts the smallest gateway that should be filled in by the nodes for successful verifications. Because the significance of validators is established in light of preceding successful verifications, a PoI protocol ensures greater confidence among contributing nodes. Thus, this protocol may be helpful for AI appliances in the general blockchain. However, in a consortium and private networks, PoI could lead to significant conflicts among shareholders since key shareholders could monopolize the entire system.

6.6 Summary

This chapter provided a comprehensive taxonomy discussion of usual blockchain functions based on the distribution of AI methods and procedures, blockchain types and infrastructure, and the basic consensus methods used for decentralized distributed transaction verifications over crucial networks. Industrial internet of things (IIoT) systems encounter many challenges such as weak mobility, diversity and resource constraints, security and confidentiality vulnerabilities. Blockchain techniques can stabilize IIoT systems based on advanced interoperability, in addition to enhanced privacy and security. Moreover, blockchain can improve the reliability and scalability of the IIoT system. The mixture of blockchain accompanied by IIoT is called BIIoT, which will be discussed in the next chapter.

References

[1] Ye D, Zhang M and Vasilakos A V 2017 A survey of self-organization mechanisms in multi-agent systems *IEEE Trans. Syst. Man Cybern.* **47** 441–61

[2] Rizk Y, Awad M and Tunstel E W 2018 Decision making in multi-agent systems: a survey *IEEE Trans. Cogn. Devel. Syst.* **10** 514–29

[3] Fioretto F, Pontelli E and Yeoh W 2018 Distributed constraint optimization problems and applications: a survey *J. Artif. Intell. Res.* **61** 623–98

[4] Rehman M H U, Liew C S, Wah T Y and Khan M K 2017 Towards next-generation heterogeneous mobile data stream mining applications: opportunities, challenges, and future research directions *J. Netw. Comput. Appl.* **79** 1–24

[5] Rehman M H U, Batool A, Liew C S and Teh Y-W *et al* 2017 Execution models for mobile data analytics *IT Prof.* **19** 24–30

[6] Bottou L, Curtis F E and Nocedal J 2018 Optimization methods for large-scale machine learning *SIAM Rev.* **60** 223–311
[7] Contreras-Cruz M A, Lopez-Perez J J and Ayala-Ramirez V 2017 Distributed path planning for multi-robot teams based on artificial bee colony *2017 IEEE Congress on IEEE in Evolutionary Computation (CEC)* 541–8
[8] van Zelst S J, van Dongen B F and van der Aalst W M 2018 Event stream-based process discovery using abstract representations *Knowl. Inf. Syst.* **54** 407–35
[9] Lu H, Li Y, Chen M, Kim H and Serikawa S 2018 Brain intelligence: go beyond artificial intelligence *Mob. Netw. Appl.* **23** 368–75
[10] Kurtulmus A B and Daniel K 2018 Trustless machine learning contracts; evaluating and exchanging machine learning models on the ethereum blockchain arXiv preprint arXiv:1802.10185
[11] Kim H, Park J, Bennis M and Kim S-L 2018 On-device federated learning via blockchain and its latency analysis arXiv preprint arXiv:1808.03949
[12] Hatem M, Burns E and Ruml W 2018 Solving large problems with the heuristic search: General-purpose parallel external-memory search *J. Artif. Intell. Res.* **62** 233–68
[13] Banerjee S, Singh P K and Bajpai J 2018 A comparative study on decisionmaking capability between human and artificial intelligence *Nature Inspired Computing* (Berlin: Springer) pp 203–10
[14] Vo H T, Kundu A and Mohania M K 2018 Research directions in blockchain data management and analytics *EDBT* 445–8
[15] Lai L and Suda N 2018 Rethinking machine learning development and deployment for edge devices arXiv preprint arXiv:1806.07846
[16] Dinh T T A, Wang J, Chen G, Liu R, Ooi B C and Tan K-L 2017 Blockbench: a framework for analyzing private blockchains *Proc. of the 2017 ACM Int. Conf. on Management of Data* (New York: ACM) pp 1085–100
[17] Li Z, Kang J, Yu R, Ye D, Deng Q and Zhang Y 2018 Consortium blockchain for secure energy trading in the industrial internet of things *IEEE Trans. Ind. Inf.* **14** 3690–700
[18] Joshi D 2017 IBM, Amazon & Microsoft are offering their blockchain technology as a service *Business Insider* Oct 24
[19] Hwang G-H, Chen P-H, Lu C-H, Chiu C, Lin H-C and Jheng A-J 2018 Infinitechain: a multi-chain architecture with distributed auditing of sidechains for public blockchains *Int. Conf. on Blockchain* (Berlin: Springer) pp 47–60
[20] Androulaki E *et al* 2018 Hyperledger fabric: a distributed operating system for permissioned blockchains *Proc. of the 13th EuroSys Conf.* (New York: ACM) p 30
[21] Chen L, Xu L, Shah N, Gao Z, Lu Y and Shi W 2017 On security analysis of proof-of-elapsed-time (PoET) *Stabilization, Safety, and Security of Distributed Systems* SSS 2017 *Lecture Notes in Computer Science* vol 10616 ed P Spirakis and P Tsigas (Cham: Springer)
[22] Angelis S D, Aniello L, Baldoni R, Lombardi F, Margheri A and Sassone V 2018 Pbft vs proof-of-authority: applying the cap theorem to permissioned blockchain *Italian Conf. on Cyber Security (ITASEC) (Milan)*

Chapter 7

Combination of blockchain and IIOT

Internet of Things (IoT) is changing the current industry into a data directed smart decision-making industry. The essential characteristics of IoT pose several challenges, for example, security and privacy vulnerabilities, low mobility, and decentralization. Blockchain technology carries chances to face the challenges of IoT. This chapter investigates the combination of blockchain with IIoT. It names this mixture of blockchain together with IIoT as Blockchain of Industrial Internet of Things (BIIoT).

7.1 Opportunities for integrating blockchain with IIoT

IIoT systems face several challenges, for example, security and privacy vulnerabilities, resource constraints on IIoT devices, low mobility and the diversity of IIoT systems. Blockchain technologies could fill IIoT outlines with improved functionality and enhanced security and privacy. Furthermore, blockchain could improve the scalability and reliability of IIoT systems [1]. In short, we name the combination of blockchain with IIoT as BIIoT. Unlike current IIoT systems, BIIoT has the following possible advantages, as shown in figure 7.1.

- **Improved functionality of IIoT systems:** Blockchain could enhance the functionality of IIoT systems through changing and recording IIoT data into the blockchain. During this process, multiple types of IIoT data are compressed, extracted, modified, processed and lastly stored in the blockchain. Furthermore, the blockchain launched in the uppermost of a peer-to-peer network helps global internet usage, making it easy to operate and pass across various fragmented networks.
- **Enhanced IIoT schemes' safety:** IIoT information can be protected through blockchain because this information is encrypted with a signed through encryption/decryption key (for example, RSA, ECDSA algorithm) stored as blockchain transactions [2]. Furthermore, the combination of IIoT systems with blockchain technologies (such as smart contracts) will assist in

doi:10.1088/978-0-7503-3663-5ch7

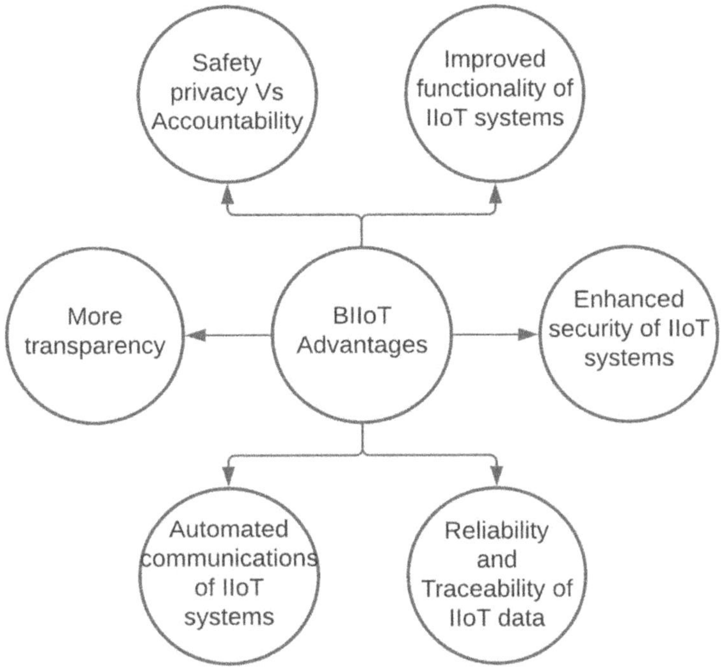

Figure 7.1. Possible advantages of BIIoT.

enhancing the safety of IIoT systems by automatically updating software for IIoT devices, thereby improving system safety to fix vulnerabilities [3].

- **Reliability and Traceability of IIoT data:** Blockchain data could be recognized and proved anywhere at any time. Concurrently, all historical transactions recorded on the blockchain are detectable. For instance, Lu and Xu [4] have grown a blockchain-based product innovation system that presents observable services to retailers and suppliers. By this method, originality and excellence of the goods can also be examined and confirmed. Furthermore, the unchanging nature of blockchain also ensures the dependability of IIoT data, as any transaction stored on the blockchain cannot be falsified or altered. Because of the difficulty of the communications between the stakeholders in a conventional supply chain, it is most challenging to precisely trace data and connect it to the object without any bias between the various stakeholders. The ability to find in the global supply chain, food, clinical trials, drugs etc is most significant. In blockchain-enabled IIoT, digital assets can be defined by a trackable resource unit (TRU), a set of one or more objects that cannot be tracked separately. Availability of details of transactions in IIoT containing certification, quality, location, date, price or other related data in the smart contract will enlarge the ability to detect the object through the blockchain chain. Blockchain can provide accurate detection of products within a complex supply chain. It can provide a worldwide detection solution.

- **Automated communications of IIoT systems:** Blockchain technologies can enable IIoT devices to communicate with each other automatically. For example, Zhang and Wen [5] suggest decentralized autonomous organizations (DAOs) for mechanizing dealings, that have no conventional characters as institutions associated in the company of fees. Developed by smart contracts, DAOs could operate spontaneously in the absence of human interference, resulting in cost savings.
- **More transparency**: Blockchain technologies make data transfer, particularly transactional data transfer, more transparent. Like a decentralized ledger, all network participants distribute similar documents as against individual copies, which can only be updated by consensus and must be agreed upon by each participant. Any change in a single transaction record will need the transformation of all ensuing forms with the agreement of the whole network. Consequently, blockchain data is used by all participants and is much more precise, stable, and transparent than that used via paper-heavy procedures.
- **Safety privacy versus accountability:** At IIoT, accountability builds trust in IoT services and IoT devices, including all transactional events; information must be eventually attributable to some agent or user. Responsibility places greater authority on users and service providers or third-party officials who hold services in charge of their operation and performance. In the blockchain, the digital pseudonym (hash of the address) is utilized to offer partial anonymity. Also, every participant in the network can discover a company's behaviours with a specific ID, and this asset also participates in making blockchain a more exciting tool to construct a tamper-proof log to be utilized in accountable IIoT [6]. Distributed ledger blockchain permits a safe platform, shared, uninterrupted, and needing a large enough network of unreliable peers. The secrecy in blockchain-enabled IIoT contributes fine-grained access control, data audit ability and transparency, data ownership, containing permissions of IoT devices and data in IIoT.

7.2 IIoT with blockchain combination (BIIoT)

IIoT is also enhancing physical procedures to make them a fraction of a digital epoch, getting quantities of information that present a previously undeveloped skill. This skill enables the growth of smart industries; for example, the development of administration as well as the excellence of the existence of machines by digitalizing services in industries. Recently, cloud computing offered IIoT functionality to examine and execute data and turn it into actual-time knowledge and actions [7]. The unfamiliar development of IIoT has initiated new social changes, for example, mechanisms for accessing and distributing data. An open information model is at the forefront of these efforts. A significant implication of these efforts is, as has happened in numerous situations, an insufficiency of trust. Federal structures such as those utilized in cloud computing contributed notably towards the growth of IIoT. They perform like a black box on data transparency and do not provide plain sight of how or where data they present for network participants is utilized.

Combining promising technologies, such as cloud computing and IIoT, has proven extremely useful [8]. Similarly, we accept the immense possibility of blockchain in transforming IIoT. Blockchain could enhance IIoT by presenting a reliable distributing service where data is dependable and can be detectable. Data sources can always be revealed and the data will remain unchanged over time, increasing its safety. If IIoT data needs to be distributed safely between multiple participants, this combination will mark a significant revolution. For example, the comprehensive discoverability of many food items is an essential feature of ensuring food security. Food traceability may need numerous participants' participation: production, treatment, feeding, sharing, and more. Data leakage in any part of the chain can lead to cheating and limit the search procedures for contamination, which could severely impact the lives of citizens and cause substantial financial costs for countries, departments with companies in cases of foodborne contamination. Ultimate control over these areas will enhance food security, improve data distribution among participants, and decrease likelihood of foodborne discharge, saving human lives.

Furthermore, in different fields, for example, smart cars and smart cities, reliable data distribution can be beneficial in recruiting novel contributors to environmental organization and contributing to the improvement of the adoption of their services. Thus, the usage of blockchain could fill IIoT with reliable and safe data. It is beginning to be identified, where blockchain technology is recognized as a critical component for resolving scaling, confidentiality and dependability issues associated with the IIoT model.

IIoT could gain considerable advantage from the functionality presented by blockchain and also grow present IIoT technologies. There are still numerous open problems and research challenges to explore in using these two technologies smoothly and continuously in conjunction, although this study area is still in its infancy.

More precisely, this incorporation may contain (though not exclusively) the following advances:

- **Scalability and decentralization:** Switching from a centralized structure to a P2P shared location eliminates focal points for failure and interruptions. It would assist in avoiding scenarios where a small number of strong companies regulate the working and records of vast numbers of people's data. Different advantages that come with architectural deployment are the enhancement of fault tolerance and computer scaling. This will decrease IIoT silos and further participate in enhancing IIoT accessibility.
- **Identity:** Participants can recognize each device utilizing a general blockchain system. The data supplied and delivered to the system is unchanged and uniquely discovers the real data presented by a machine. Furthermore, blockchain can offer reliable shared authentication and device authorization for applications of IIoT [9]. It marks progress in the IIoT industry with its contributors.
- **Self-sufficiency**: Blockchain enhances the next-generation appliance characteristics that can create smart, independent resources with hardware-as-a-

service (HaaS). With blockchain, devices are competent in communicating with each other without any servers' participation. IIoT applications can gain from this operability for being device-aware with disconnected appliances.

- **Dependability:** IIoT data will remain unchanged and will share in blockchain over time. The system participants are competent in checking the data's reliability and are sure that it is not corrupted. Furthermore, their skill allows sensor information detection with responsibility. A vital feature of overcoming a barrier for adopting IIoT is reliability [10].
- **Security:** Communications and data can be protected if they are recorded as blockchain transactions. Blockchain can be considered as device messaging as transactions are verified by smart contracts, thus protecting communications between machines. Recent safe standard-protocols utilized in IIoT could be improved by the use of blockchain [11].
- **Services market:** Blockchain could speed up a formation of the IIoT environmental organization of services with information marketplaces where dealings among peers are feasible in the absence of authorities. Microservices can easily be utilized, and micropayments can be created securely in reliable surroundings. This will enhance IIoT interconnection and use IIoT data on the blockchain.
- **Deployment of safe code:** Using blockchain safe-unchanged storage, the code can be safely shared across devices. Producers can trace states and updates with greater trust. IIoT middleware can utilize this function to update IIoT devices safely.

Another feature to consider is the relationship between IIoT communications, for example, the interaction between necessary IIoT infrastructures. When combining blockchain, it is essential to determine where these communications occur: within IIoT, the mixture plan concerning blockchain with IIoT, or via blockchain. The novel layer among IIoT devices and the cloud has revolutionized Fog computing, making this integration easier.

In conventional IIoT deployments, restricted-asset devices utilized like endpoints to be contacted by the entrance are accountable for transmitting sensor information for the higher layers. While combining blockchain, the latter devices can provide cryptographic practicality on IIoT devices to communicate with the blockchain. It is essential to understand IIoT autonomy, which comes with very sophisticated hardware with a high computational cost. Combining gateways into these solutions is further reliable, as are conventional deployments, although the advantages of utilizing blockchain for this method are minimal. Regardless of the growth of blockchain, it makes no sense to exploit it in a lot of appliances where a database presents good practicality. Determining when to utilize a blockchain depends primarily on the needs of a user. For example, where the highest level of effectiveness is needed, blockchain might not only be the best solution; however, the mixture method can be used to improve it.

Alliances between familiar companies have begun to seem, as the trusted IIoT alliance, to sustain the links between blockchain and IIoT [12]. Several devices with

Table 7.1. IIoT devices for utilized like blockchain elements.

Source	IIoT device	Mode	Blockchain
Bitmain	Antrouter R1-LTC	Miner	Litecoin
Raspnode	Raspberry Pi	Full node	Bitcoin Ethereum Litecoin
Ethraspbian	Raspberry Pi	Full node	Bitcoin Ethereum Litecoin
EthEmbedded	Raspberry Pi BeagleBone Black Odroid XU3/XU4 Wandboard Ethcore Parity	Full node	Ethereum

combined blockchain abilities also exist in the market. EthEmbedded permits Ethereum to install full nodes in embedded devices, for example, Odroid, Raspberry Pi and Beaglebone Black [13]. Both Ethraspian and Raspberry assist Ethereum, and Bitcoin, Litecoin full nodes in a Raspberry Pi. The Antrouter R1-LTC is a Wi-Fi router which further activates mining Litecoin [14]. Thus, this kind of router can be established in smart homes and is a fraction of the fog computing environmental organization. Raspnode supports wallets for Litecoin and Bitcoin. As Raspnode states, mining can be completed on IIoT devices. However, it is purposeless. Mining has become the unique work for some hardware; moreover, it is senseless to attempt this in IIoT devices. It is a primary reason why there is little activity in mining for IIoT devices. Much more research is needed to integrate IIoT devices widely into blockchain elements. Table 7.1 demonstrates an outline of the reviewed IIoT devices utilized as part of the blockchain platforms.

Full nodes should record the whole blockchain (recently between 46 to 150 GB at Ethereum and Bitcoin, correspondingly) for complete verification of transactions and blocks. Therefore, their ordering is more restricted on IIoT devices. As mentioned, the mine will be useless in IIoT because of its particular necessities. Consensus protocol can relax to facilitate IIoT devices' inclusion, but this could compromise the blockchain execution's safety. It can be utilized by a consortium blockchain where consensus protocol is settled. For exploitation of lightweight devices, the legitimacy of transactions is authenticated with no requirement to download a whole blockchain; thus, they can participate in blockchain, which is simple for execution and can also be kept in IIoT devices. These devices can be utilized in IIoT to be a fraction of a blockchain system, decreasing an interval between two techniques. It always has to be backed up with full nodes for authenticating blocks and dealings. Nonetheless, much of the blockchain does not support lightweight devices as is the case of Ethereum, where is is under development. Moreover, blockchain can utilize an outside service for security and dependable storage.

One obvious substitute for blockchain combination with IIoT is a combination of cloud computing and IIoT. This combination has been utilized to conquer IIoT restrictions over the past few years: executing, record of use. Cloud computing typically presents the federal framework that, different from blockchain, makes dependable distributing difficult with multiple contributors. A combination of IIoT and blockchain is planned to tackle the aforementioned restrictions as well as keeping reliable information. Fog computing follows a shared method, for example, blockchain, which aims to share computing closer to the end devices [15]. It can connect more powerful devices than IIoT such as edge nodes and gateways, which can then be used again as blockchain elements. Thus, fog computing will make it easier to integrate blockchain with IIoT.

7.3 Challenges at a blockchain–IIoT incorporation

Significant challenges to be tackled when applying blockchain to an IIoT area are discussed in this section. Blockchain is planned for the internet situation by useful computers, and this is distinct from an IIoT actuality. The blockchain transactions are signed digitally; therefore, nodes competent to work with money should be prepared by this operability. Incorporating blockchain in an IIoT is demanding. This section presents a few of the recognized challenges.

7.3.1 Capability and scalability of storage

Storage capability with the scalability of blockchain is discussed below; however, in the case of IIoT appliances, the intrinsic ability with scalability restrictions makes these challenges very important. In this sense, blockchain can emerge as inappropriate for IIoT machines, but there are methods in which these restrictions can be completely avoided. In IIoT, where devices can create gigabytes of information in an instant, this restriction signifies a massive obstruction for its combination with the blockchain. It is widely recognized that blockchain executions can only conduct a few transactions per second; thus, this could present a possible traffic jam for IIoT. Moreover, blockchain is not intended to record a massive quantity of data similar to that created in the IIoT. A combination of these technologies must tackle these challenges.

Much IIoT data is recorded, only a part of which is helpful for providing information to carry out actions. Various existing methods to compress, regularize and filter IIoT data with the aim of reducing it have been proposed. The IIoT is engaged with target and communication services (cloud, blockchain), and embedded devices; therefore, savings in the quantity of information that the IIoT presents could benefit numerous layers. Data compression could ease broadcast, processing jobs also are records of a massive number of IIoT details created. Usual activities do not typically need more than essential data, not like irregular information.

Finally, blockchain, especially its consensus protocol that affects its traffic jam, furthermore is modified to enlarge bandwidth and reduce delay of its dealings, thus allowing an enhanced changeover to the IIoT established through the Bitcoin-NG case [16].

7.3.2 Safety

IIoT appliances tackle safety troubles in various stages, however, with an extra difficulty because of performance with many devices' heterogeneity. Additionally, the IIoT circumstance consists of a collection of property that causes safety issues, for example, mobility, scale or wireless communication.

The growing quantity of attacks on IIoT networks, and their severe effects, make it even more essential to make an IIoT with very complicated safety measures. Numerous specialists observe blockchain as the critical skill to offer a significantly required safety enhancement in IIoT. One of the significant challenges in linking IIoT to a blockchain is the dependability of information created through IIoT. Blockchain can guarantee that information at a chain is unchanged and can recognize its alteration. However, when information arrives previously corrupted at a blockchain, it remains in that state. Dishonest IIoT information could occur from numerous conditions apart from spiteful ones. The welfare of an IIoT structural design can be caused through multiple elements, for example, surroundings, contributor, and damage as well as device breakdown. Occasionally devices themselves or their actuators and sensors fail for a correct job from the start. This circumstance cannot be distinguished until a machine query has been examined, otherwise occasionally it works precisely while changing its performance for numerous purposes. Additionally to these circumstances, innumerable threats could occur for the IIoT, such as eavesdropping and denial of service (DoS) [17]. Therefore, IIoT devices must be methodically tested before their combination with blockchain, and they must be situated with encapsulating at the correct position to avoid physical damage, as well as containing methods to identify node breakdowns immediately they occur.

These nodes are very susceptible to hacking because their restrictions regulate a firmware update, preventing them from actuating more than reasonable errors. Furthermore, it is in some cases hard to update nodes separately, like in worldwide IIoT deployments. Thus, run-time improvement with reconfiguration method must be located in the IIoT to maintain its continuous running. Initiatives such as REMOWARE and GUITAR that allow system and firmware updates in run-time are necessary to guarantee a safe combination of the IIoT with blockchain [18].

The blockchain and IIoT combination could hold further consequences for IIoT interactions. Presently, IIoT application protocols, for example, MQTT and CoAP utilize different safety protocols, such as DTLS or TLS to present safe communications [19, 20]. These security protocols are complicated and need centralized management, as well as the supremacy of crucial communications, usually by public key cryptography (PKI). At a blockchain network, every IIoT node will contain its own global unique identifier or pair of asymmetric keys installed once linked with a network [21]. It will make it simpler to present safety protocols which are typically designed to swap PKI certificates and will permit them to be utilized in devices with inferior ability.

One noteworthy IIoT scheme about safety with the blockchain acceptance is filament. The filament is the software with hardware result which presents payments functionality based on bitcoin and smart contracts at IIoT. Filament nodes contain embedded crypto processors which sustain five protocols: Smart contracts, Telehash with Blockname to function, also Bittorrent and Pennyback protocols. A node individuality administration is completed with block name. Simultaneously, Telehash, the free execution of Kademlia DHT, presents safely encrypted interactions; also, smart contracts describe the method within the node that can be utilized.

7.3.3 Information confidentiality with anonymity

For numerous IIoT appliances tasked with secret information, for example, while a node is connected to an individual, for example in the eHealth situation, it is necessary to deal with the difficulty of information anonymity and privacy. Blockchain is provided as a perfect outcome for tackling distinctiveness management with IIoT, such as with bitcoin, there might be appliances where secrecy need to be ensured. These are wearable devices with the capability to cover an individual's anonymity while transmitting private information. These otherwise smart vehicles defend solitude of an itinerary of customers.

The difficulty of information confidentiality in a prominent public blockchain has previously been argued, collectively by a few of the previous solutions. But, the complication of information confidentiality at IIoT nodes involves much complexity, because it initiates in information planned to increase interaction and appliance stages. Protecting a node, which data are stored safely and not used by citizens in the absence of permission is confronted, needs the combination of safety cryptographic software within a node. These enhancements must encompass a restriction of assets of nodes and limitations associated with financial feasibility. Numerous technologies contain safe interactions based on encryption (DTLS, SSL/TLS, IPsec). IIoT device restrictions frequently make it essential to utilize fewer controlled nodes, for example, gateways to integrate these safety protocols. Use of cryptographic hardware can speed up the cryptographic process and avoid complicated safe software protocols.

Security of information with confidentiality is a core problem for IIoT; utilizing blockchain can improve the difficulty of ID management in IIoT. Trust is another essential characteristic of an IIoT where a combination of blockchain could run the task. Data truthfulness technique is another choice to guarantee data usage simultaneously as it avoids congestion of blockchain by the enormous quantity of data created by the IIoT. It could be an outcome in public systems, however, with a proficient and limited access control. MuR-DPA presents energetic data update and professional confirmation through public auditing confirmation [22].

Finally, some laws control data confidentiality; for example, the EU's data security commands will require changes to cover a new form which a skill makes potential. An acceptance of making blockchain-like the lawful stage must tackle these rules, thus guaranteeing information confidentiality following a law.

7.3.4 Smart contract

IIoT can benefit from a smart contract, but how it fits within the IIoT appliance is varied. Since the realistic viewpoint, a deal is a set of data (states) and code (functions) that exist in an exact blockchain address. Public tasks at the contract could call through devices. Functions could further fire events; appliances could pay attention to them correctly respond to the function fired. A transaction must be published on the network to change the contract's status, in particular, to change the blockchain. Transactions signed through senders also contain to be accepted by the network.

The IIoT can sense and actuate over the internet in numerous fields. In the case of food trackability, food packaging is equipped with sensors with the ability to calculate ecological circumstances and attach them to a blockchain. In the blockchain, the contract will offer operations to start and stop delivery as well as registering and questioning quantities. When quantities surpass the pre-defined threshold, action will be triggered. The managing appliance will pay attention to these actions, and shipping companies, retailers, producers with customers will be notified. If no proceedings are increased, a blockchain would ensure that a consignment is carried out in the most favourable circumstances. Smart contracts will also present a dependable executing engine for an IIoT, registration again running all their communications. Action will be an outcome of the trustworthiness with safe execution. Thus, smart contracts can be a safe method for the operation of IIoT appliances. However, a subsequent challenge must be tackled with that combination.

Functioning with smart contracts needs oracles, a particular entity that gives actual-world information for the reliance method. Authenticating this smart contract can be compromised as the IIoT might be not stable. Currently, smart contracts are disseminated and decentralized. However, they do not perform to distribute assets to allocate jobs and tackle the massive quantity of calculation. The implementation of a smart contract is completed on a solitary device while concurrently a code implementation is achieved through numerous nodes. This sharing only ends at the confirmation procedure, rather than utilizing it to share jobs. An IIoT encourages the shared abilities of big data with cloud computing to raise its executing energy. Thus, data mining methods are capable of tackling IIoT information as a whole, allowing the superior comprehension of an IIoT; for example, executing energy extended through cloud computing. Big data have allowed the execution of massive amounts of information concurrently, permitting extraction from massive datasets, which was formerly extremely tough to perform. With a mixture of IIoT with blockchain, smart contracts must influence their disseminated environment to allow an implementing ability offered by various models and wanted in the IIoT.

Smart contracts must take into account the restrictions and heterogeneity provided in the IIoT. Group mechanisms and filtering must be complemented through smart contracts to allow appliances to tackle an IIoT relying on a circumstance with necessities. The detection method can allow device inclusion on the fly, making this application very powerful. Finally, actuation methods directly from smart contracts will allow a quicker response by the IIoT.

7.3.5 Legal problems

An idea of the non-controled blockchain is a fraction of its quintessence, and partially responsible for bitcoin's success. Blockchain, particularly in the case of virtual money, has carried numerous controversies concerning legitimacy. The necessity, or chance, to initiate control components over the network has arrived with the structure of the consortium, private and permissioned blockchain.

An IIoT field is affected by a country's rules concerning information confidentiality, for example, information security instruction. Many of these rules become outdated and require changing, particularly with the appearance of new disrupting technologies such as blockchain. The growth of the latest practices and principles could ease the certification of safety features of devices, and with this method construct the safest and dependable IIoT network. In this logic, rules that deal with data confidentiality and data management that are still a considerable challenge to begin with IIoT will thus be an even more enormous challenge if used in the mixture with blockchain.

The need for regulation has drawbacks for the method for private key reset or retrieval. Otherwise, transaction deterioration is not feasible. A few IIoT appliances visualize a worldwide, distinctive blockchain for nodes. But it is unclear if this kind of network is controlled through producers or unlocked for the consumer. Moreover, it is anticipated to need legal ruling. These rules would have authority over the future of blockchain with IIoT and as such could probably disturb the decentralization and open environment of blockchain by initiating a regulating, centralized contributor, for example, a country.

7.3.6 Consensus

In IIoT appliances, the restricted asset environment of nodes makes them inappropriate for capturing parts of consensus methods, for example, proof of work (PoW), openly. There is a broad diversity of suggestion for consensus protocols, but they are usually undeveloped and have not been examined sufficiently. Asset necessities rely on an exacting kind of consensus protocol in a blockchain network. In general, the solution tends to reserve these jobs for gateways, or any other unrestrained node, proficient at presenting this operability. Optionally off-chain results, that shift data outside a blockchain to reduce any substantial delay in blockchain, can give operability.

Though there is the plan to integrate blockchain full nodes within IIoT nodes, mining is still the core problem in an IIoT because of its restrictions. IIoT is mostly composed of asset-restricted nodes; however, worldwide IIoT has potentially enormous implementing energy, as captured in a report that anticipated the number of nodes would reach anywhere between 20 to 50 billion by 2020. Study attempts must be spotlighted in this domain that leverage for a dispersed environment worldwide the possibility of an IIoT to modify a consensus in an IIoT.

At Babelchain, the new consensus protocol named Proof of Understanding (PoU) is proposed, which aspires to adapt PoW for IIoT appliances [23]. With less power usage, a protocol, rather than utilizing miners to resolve hash puzzles, recommends translating various protocols. An attempt is very focused on helpful calculation though concurrently dealing with a critical difficulty in IIoT interactions. Peers in the network, rather than approving on transaction rank, accept on message meaning (action, content and format). Moreover, the blockchain information gives data to study, similar to a learning set.

7.4 BIIoT usage

The blockchain technology is broadly utilized in industries including economics, education, real estate, power, e-healthcare, manufacturing, and e-government. In economics, blockchain demonstrates enormous promise and has been utilized to construct novel investment and payment systems and cryptocurrency.

In e-government systems, the blockchain can present effectual identification confirmation over a disseminated ledger to store identity, for example, blockchain-based land registration, e-identity in Estonia, e-passport in Dubai, and so on. New supply chains have become more complicated; their consequences for numerous companies' competitiveness is a significant aspect to take into account.

In blockchain-enabled IIoT appliances, Ethereum is one of the most well-liked platforms, which could present more features. The rising 5G technology and M2M could considerably improve the exploitation of blockchain in IIoT [24].

7.5 Summary

This chapter investigated the combination of blockchain accompanied by IIoT. It names a mixture of blockchain with IIoT as blockchain of industrial internet of things (BIIoT). This chapter also discussed opportunities for integrating blockchain with IIoT, challenges in blockchain–IIoT integration and usages of BIIoT. The architecture of BIIoT will be discussed in the next chapter.

References

[1] Reyna A, Martn C, Chen J, Soler E and Daz M 2018 On blockchain and its integration with IoT. Challenges and opportunities *Future Gener. Comput. Syst.* **88** 173–90

[2] Farooq S M, Hussain S S and Ustun T S 2019 Elliptic curve digital signature algorithm (ECDSA) certificate-based authentication scheme for advanced metering infrastructure *2019 Innovations in Power and Advanced Computing Technologies (i-PACT)* vol 1 (Piscataway, NJ: IEEE) pp 1–6

[3] Fotiou N, Siris V A and Polyzos G C 2018 Interacting with the Internet of Things using smart contracts and blockchain technologies *Int. Conf. on Security, Privacy and Anonymity in Computation, Communication and Storage* (Cham: Springer) pp 443–52
[4] Lu Q and Xu X 2017 Adaptable blockchain-based systems: a case study for product traceability *IEEE Softw.* **34** 21–7
[5] Zhang Y and Wen J 2017 The IoT electric business model: using blockchain technology for the internet of things *Peer Peer Netw. Appl.* **10** 983–94
[6] Feng Q, He D, Zeadally S, Khan M K and Kumar N 2019 A survey on privacy protection in the blockchain system *J. Netw. Comput. Appl.* **126** 45–58
[7] Li Q, Yue Y and Wang Z 2020 Deep robust cramer shoup delay optimized fully homomorphic for IIOT secured transmission in cloud computing *Comput. Commun.* **161** 10–8
[8] Han G, Que W, Jia G and Zhang W 2018 Resource-utilization-aware energy-efficient server consolidation algorithm for green computing in IIOT *J. Netw. Comput. Appl.* **103** 205–14
[9] Hammi M T, Hammi B, Bellot P and Serhrouchni A 2018 Bubbles of trust: a decentralized blockchain-based authentication system for IoT *Comput. Secur.* **78** 126–42
[10] Liu S, Guo C, Al-Turjman F, Muhammad K and de Albuquerque V H 2020 Reliability of response region: a novel mechanism in visual tracking by edge computing for IIoT environments *Mech. Syst. Sig. Process.* **138** 106537
[11] Sengupta J, Ruj S and Bit S D 2020 A comprehensive survey on attacks, security issues and blockchain solutions for IoT and IIoT *J. Netw. Comput. Appl.* **149** 102481
[12] Narayanan R and Murthy C S 2017 A probabilistic framework for protocol conversions in IIoT networks with heterogeneous gateways *IEEE Commun. Lett.* **21** 2456–9
[13] Afanasev M Y, Fedosov Y V, Krylova A A and Shorokhov S A 2018 An application of blockchain and smart contracts for machine-to-machine communications in cyber-physical production systems *2018 IEEE Industrial Cyber-Physical Systems (ICPS)* (Piscataway, NJ: IEEE) pp 13–9
[14] Waheed A and Krishna P V 2020 Comparing biometric and blockchain security mechanisms in smart parking system *2020 Int. Conf. on Inventive Computation Technologies (ICICT)* (Piscataway, NJ: IEEE) pp 634–8
[15] Kochovski P, Gec S, Stankovski V, Bajec M and Drobintsev P D 2019 Trust management in a blockchain-based fog computing platform with trustless smart oracles *Future Gener. Comput. Syst.* **101** 747–59
[16] Reyna A, Martín C, Chen J, Soler E and Díaz M 2018 On blockchain and its integration with IoT. Challenges and opportunities *Future Gener. Comput. Syst.* **88** 173–90
[17] Tajalli S Z, Mardaneh M, Taherian-Fard E, Izadian A, Kavousi-Fard A, Dabbaghjamanesh M and Niknam T 2020 DoS-resilient distributed optimal scheduling in a fog supporting IIoT-based smart microgrid *IEEE Trans. Ind. Appl.* **56** 2968–77
[18] Ismail S, Almayouf R, Chehab S, Alghamdi S, Almutairi A, Alasmari B and Altherwy R 2020 Edge IoT-cloud framework based on blockchain *2020 2nd Int. Conf. on Computer and Information Sciences (ICCIS)* (Piscataway, NJ: IEEE) pp 1–7
[19] Ansari D B, Rehman A and Ali R 2018 Internet of Things (IoT) protocols: a brief exploration of MQTT and CoAP *Int. J. Comput. Appl.* **975** 8887
[20] Tiburski R T, Amaral L A, de Matos E, de Azevedo D F and Hessel F 2017 Evaluating the use of TLS and DTLS protocols in IoT middleware systems applied to E-health *2017 14th*

IEEE Annual Consumer Communications & Networking Conf. (CCNC) (Piscataway, NJ: IEEE) pp 480–5

[21] Knezevic D B and Kasunic N 2020 Security challenges of Wi-Fi connected beer cooler and serving IIoT device *2020 5th Int. Conf. on Smart and Sustainable Technologies (SpliTech)* (Piscataway, NJ: IEEE) pp 1–5

[22] Geetha B G 2018 Signature verification and bloom hashing technique for efficient cloud data storage *Wirel. Pers. Commun.* **103** 3079–97

[23] Qiu Z, Hao J, Guo Y and Zhang Y 2020 Dual vote confirmation based consensus design for blockchain integrated IoT *NOMS 2020–2020 IEEE/IFIP Network Operations and Management Symp.* (Piscataway, NJ: IEEE) pp 1–7

[24] Zhao S, Li S and Yao Y 2019 Blockchain-enabled industrial Internet of Things technology *IEEE Trans. Comput. Social Syst.* **6** 1442–53

IOP Publishing

Blockchain in the Industrial Internet of Things

Lakshmana Kumar Ramasamy and Seifedine Kadry

Chapter 8

BIIOT architecture

Blockchain is an excellent complement for IIoT with advanced scalability, reliability, security, confidentiality and mobility. Combining blockchain and an IIoT network enhances the overall safety of a network. Blockchain has the full protection with confidentiality features which the IoT enabled smart industry needs. Therefore, this chapter examines the latest model for linking IIoT to the blockchain. We name the combination of blockchain and IIoT as the blockchain of industrial internet of things (BIIoT). The structure of BIIoT is proposed in this chapter.

8.1 Architecture of BIIoT

Figure 8.1 demonstrates the architecture of the BIIoT for a smart industrial environment.

BIIoT is supple and permits delopers to change in line with industry necessities. Subsequent sections explain all the layers of BIIoT architecture in brief.

8.2 Physical layer

The physical layer includes microcomputers, actuators and sensors in smart industrial surroundings. We regard four sensors in BIIoT architecture: weight, pressure, humidity, and temperature sensors [1]. These sensors are integrated with the Ultra Power STM32 development board, which sends the sensed information to a blockchain service layer, afterwards executing a few necessary pre-processing actions. Three actuators are regarded in this architecture—robotic manipulators, solenoid valves and DC gear motors [2]. To proficiently manage and converse with these actuators by a service layer of blockchain, smart actuator drivers further outline the fraction of a physical layer.

doi:10.1088/978-0-7503-3663-5ch8

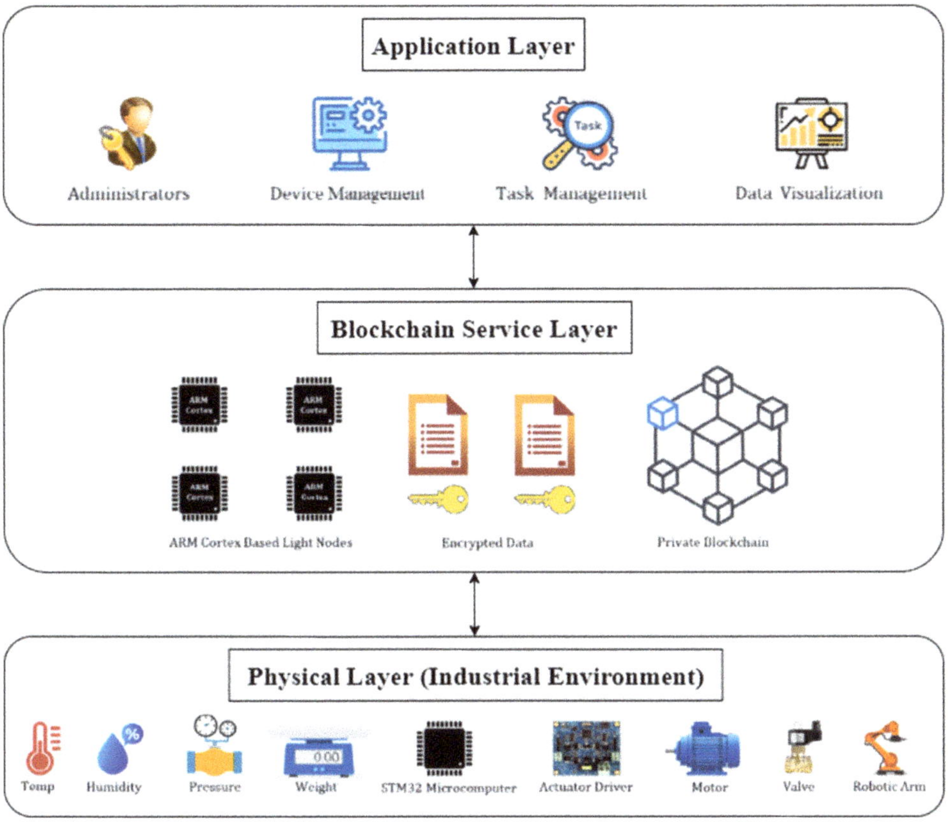

Figure 8.1. BIIoT architecture.

8.3 Service layer of blockchain

A service layer of blockchain has significant blocks that arrange the general services required to present the necessary blockchain characteristics. In BIIoT architecture, a service layer of blockchain is also sub-divided as two sections: 1) nodes with lightweight; 2) private blockchain [3].

8.3.1 Nodes with lightweight

The nodes with lightweight are not utilized for primary information recording; instead, they specially planned a rapid execution of asymmetric cryptographic procedures [4]. This is implemented using low-power ARM Cortex M4 processors containing STM32F427 development boards [5]. The entire sequence of contemporary processors is available for IoT appliances; however, the ARM Cortex-M sequence is the outstanding option for executing cryptographic procedures. The elliptic curve digital signature algorithm (ECDSA) is utilized to make private and public keys and authentication of device methods. ECDSA is a perfect option because of its ease of use and small storage necessities [6].

The STMicroelectronics *X*-CUBE-CRYPTOLIB library presents a robust platform for executing numerous typical cryptographic algorithms accompanied by the ARM Cortex-M sequence processors. A NIST Cryptographic Process Verification Program is accepted and certified by ECDSA for industrial appliances [7]. Instead of setting up the solitary node, it expands numerous nodes which gives the right of entry for a blockchain network to devices and users. Using various dispersed nodes decreases the load compared to utilizing a solitary node and prevents scheme breakdown from happening because of a solitary crashed node. Thus, this plan is an extremely significant result to integrate blockchain with actual industrial surroundings.

8.3.2 Private blockchain

The blockchain method of BIIoT is building in the private network, which presents the right of entry to permitted customers only [8]. A network of blockchain is the disseminated ledger which keeps the transactions along with process records within industrial surroundings. This ledger has the chain of cryptography-based connected blocks which are the integral fraction of a blockchain scheme. Every block has the Merkle root, hash value, previous block hash value and timestamp. Figure 8.2 illustrates the BIIoT block architecture. These blocks further allow a peer-to-peer system and information broadcast in the decentralized system.

External customers with appliances can get right of entry to the ledger and change according to their needs through utilizing a smart contract. This contract has fixed logic with code using a mathematical function. Smart contracts present customer right of entry for a blockchain system exclusive of participation of a third party through implementation alone. Smart contracts are permanent, so it is difficult for an unidentified individual to create modifications in a blockchain. Immutability is the attribute of a blockchain that does not let transactions in a ledger be changed by an unauthorized person.

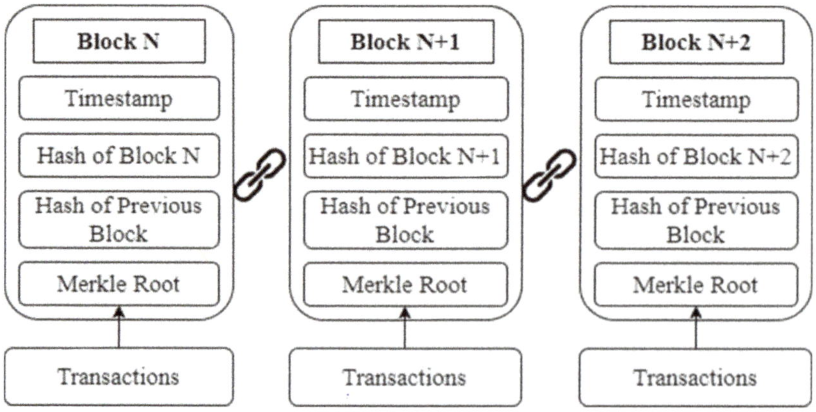

Figure 8.2. Block architecture of the proposed system.

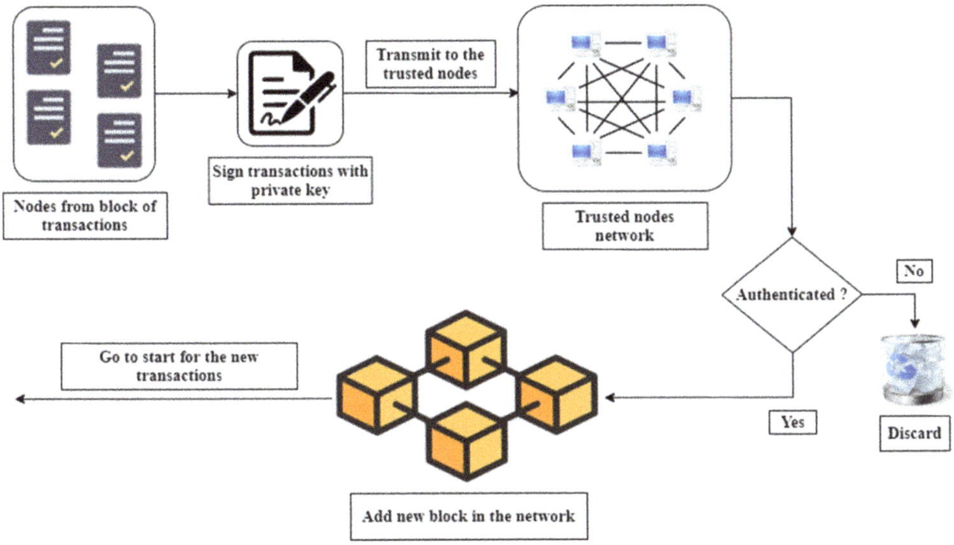

Figure 8.3. The block diagram of a PoAh consensus procedure.

8.3.3 Consensus procedure

The consensus procedure permits novel transactions to add to a blockchain. Researchers possess different consensus algorithms, for example, algorithms using proof of activity (PoA), proof of work (PoW) and so on. In BIIoT architecture, 'proof of authentication (PoAh)' is used for modernizing a blockchain [9]. Figure 8.3 demonstrates the block diagram of a PoAh consensus procedure.

At the start of the procedure, network contributors make transactions with gathered information to make the block. In a second step, nodes sign a trade by the private key and transmit to a network. Simultaneously, the block is obtained in a network after which waiting nodes discover a sender public key to verify the signature. This is followed by successful verification of the signature, trusted nodes assess a Media Access Control (MAC) tackle and compare it with an obtained one to a subsequent round of assessment [10]. After successful verification, authenticated blocks are transmitted through waiting nodes with the discovery of PoAh. Afterwards, consumers on a network check PoAh data for attaching blocks to a chain. Then accepting the suitable block, a consumer calculates the hash value for connecting a subsequent block and gets back the previous block hash value for storage within a new block. Algorithm 8.1 provides the implementation procedure of a PoAh.

Algorithm 8.1: Implementation procedure of a PoAh.

Inputs : All the nodes at the network follow ECDSA.
Outputs : Validated blocks that are attached to the blockchain.
Step 1 : Nodes merge numerous transactions to create the blocks

Step 2 : Nodes sign blocks with transmission to the network
Step 3 : The hoped node confirms the signature by the source public key
Step 4 : If (Block Authentication = = true)
Step 5 : A waiting node transmits the authenticated block to a network
Step 6 : Add this novel block to the blockchain
Step 7 : Else
Step 8 : Remove the invalid block
Step 9 : End

The PoAh method follows a conventional blockchain functioning replica by light-weight block confirmation. It removes the overturn hash function utilized in a PoW method to create a transaction procedure light in weight. Consequently, blockchain can be proficiently incorporated, accompanied by resource inhibited devices in an IIoT.

8.3.4 Application layer

The application layer presents various borders for customers who can then manage devices through imagining information utilizing physical devices. The application layer offers numerous services, including task management, user management and administration, etc.

8.4 BIIoT architecture execution procedure

The BIIoT architecture planned for allows safe function management in smart industrial surroundings. These tasks comprise system initialization, actuator and sensor information storage, or customer services which execute industrial functions in a reliable method. All these functions are described in the following subsections in brief.

8.4.1 System initialization

In the initialization stage, users with equivalent IoT nodes are recorded in a blockchain network. It is necessary to allow a re-authorization of customers with devices. A register procedure is also separated into two stages: (1) device registration (2) user registration.

8.4.1.1 User registration

At this stage, the system admin creates a sole novel ID for the customer. After that, an administrator transmits this ID for a blockchain device in the transaction suggestion. A blockchain device initially confirms the availability of this ID in a system utilizing the smart contract. If an ID was previously available, the transaction is refused, and an alert is triggered to an administrator. While an ID does not yet exist in a network, a smart contract permits a transaction. Then a PoAh method is implemented, the novel block created which is disseminated between all blockchain devices. If a customer is effectively recorded, a blockchain creates the customer ID credential utilizing its private key and transmits it for administrators. Administrators can extract certificate data using their private keys. An implementation of this procedure is also detailed in Algorithm 8.2.

Algorithm 8.2: Registration of user.

Inputs : User Id.
Outputs : Registration of User.
Step 1 : Admin creates a distinctive ID for the user
Step 2 : User ID: SHAHID314
Step 3 : if (User_ID_exist != false)
Step 4 : Refuse transaction
Step 5 : return error()
Step 6 : else
Step 7 : Apply PoAh
Step 8 : Store user ID at blockchain
Step 9 : Return user Id certificate
Step 10: End

8.4.1.2 Device registration

A physical layer has actuator and sensor nodes in BIIoT architecture. Then the necessary pre-processing is executed on the actuator's or sensor's information, a microcomputer based on STM32 transmits data for the lightweight device in a blockchain service layer and a light device processes a node register demand for a blockchain device as the novel transaction suggestion. The smart contract confirms a node ID and implements a transaction in a blockchain network. When a PoAh procedure is finished, the novel block is created and disseminated between all blockchain nodes. All the actuators and sensors are stored in a blockchain network by the unique ID. Algorithm 8.3 explains the implementation of the device registration procedure.

Algorithm 8.3: Registration of Device.

Inputs : Device Id.
Outputs : Registration of Device.
Step 1 : IoT device creates a unique ID
Step 2 : Device ID: TEMP587
Step 3 : Microcomputer transmits the data to light nodes
Step 4 : Light nodes execute the registration demand
Step 5 : if (Device_ID_exist != false)
Step 6 : Refuse transaction
Step 7 : return error()
Step 8 : else

Step 9 : Implement PoAh
Step 10 : Store device ID at blockchain
Step 11 : Provide the announcement of successful registration
Step 12 : End

8.4.1.3 Information storage

All actuators and sensors are effectively stored in the blockchain network through a primary stage. A subsequently significant fraction of function administration is registered information in a blockchain network from a held device. Now, STM32 executes the essential pre-processing in the sensor information and transmits it for the lightweight device of a blockchain service layer. Initially, a lightweight device based on ARM Cortex-M4 confirms a device through utilizing its register ID. If a blockchain validates nodes, after that a processor performs ECDSA at the actual time. This procedure encrypts information which creates the public with private keys. All the data is copied to all lightweight devices that have the backup ability so that solitary devices coincide. Following an encryption procedure, a lightweight device communicates with the smart contract to perform a transaction. A smart contract permits a transaction and records encrypted information at a blockchain network. At this time, a public key is utilized to confirm the information's reliability; also, a private key distributed only among approved consumers can decrypt information then access it from a blockchain network. All devices in a physical layer use this similar procedure. Algorithm 8.4 discusses the data storage procedure.

Algorithm 8.4: Data Storage.

Inputs : Sensor's data.
Outputs : Data storage.
Step 1 : Sensor's data are pre-processing in STM32 microcomputer
Step 2 : Transmit data to ARM Cortex light nodes
Step 3 : ARM cortex confirms the device utilizing its registration ID
Step 4 : if (device authentication != false)
Step 5 : Implement ECDSA
Step 6 : Create private and public keys
Step 7 : Submit transaction suggestion to blockchain for information storage
Step 8 : if (PoAh != false)
Step 9 : Record encrypted information in the blockchain network
Step 10 : else
Step 11 : return error()
Step 12 : else
Step 13 : Refuse transaction
Step 14 : return error()
Step 15 : End

8.4.1.4 Management of operation

BIIoT architecture makes easy the use of services of smart industry surroundings through utilizing a blockchain network. Here we examine what comprises the right of access to records of sensor information and actuator functions using their necessities.

- **Information acquisition:** To get entry to the sensor information recorded in the blockchain, a consumer presents the smart contract's transaction suggestion accompanied by correct device data. An initial smart contract checks a consumer's ID; when confirmation is successful, it then requires a consumer's public key. A consumer presents their public key; once proven valid, a blockchain would allow the right of entry to the consumer. From then, that consumer can have right of entry information from an equivalent device and decrypt it utilizing the private key. Algorithm 8.5 discusses the information acquisition procedure.

Algorithm 8.5: Information acquisition.

Inputs : Transaction proposal with device data.
Outputs : Information acquisition.
Step 1 : The consumer presents the proposal of transaction accompanied by device data
Step 2 : Blockchain nodes confirm the device and user IDs
Step 3 : if (verification != false)
Step 4 :   Request public key from the consumer
Step 5 : else
Step 6 :   return error()
Step 7 :   The consumer presents the public key of the equivalent device
Step 8 :   Blockchain network confirms digital signatures of the equivalent record
Step 9 :   if (verification != false)
Step 10 :   Implement transaction
Step 11 :   Present right of entry to the blockchain record
Step 12 :   else
Step 13 :   Refuse transaction
Step 14 :   return error()
Step 15 : End

- **Actuation job:** To manage actuators, for example, a robotic arm, solenoid valve, or motor, a consumer provides a transaction suggestion with relevant node data. With confirmation of a consumer's identity and node identity, the smart contract would permit a transaction to be performed. After successful PoAh, a blockchain would next let a consumer work an equivalent actuator. After an actuating procedure, an actuator transmits the most recent data for a blockchain service layer accompanied by an actuation job outcome. Lightweight devices will modernize blockchain storage. Algorithm 8.6 discusses an actuation procedure.

Algorithm 8.6: Actuation process.

Inputs : Transaction proposal with device data.
Outputs : Actuation process.
Step 1 : The consumer submits a transaction proposal with actuator data
Step 2 : Blockchain nodes confirm the actuator and consumer data
Step 3 : if (user_ID && actuator_ID != false)
Step 4 : Implement transaction
Step 5 : Let the consumer work the equivalent actuator
Step 6 : else
Step 7 : return error()
Step 8 : The consumer executes preferred actuation process
Step 9 : After job completion, actuator transmits the newest data to light nodes
Step 10 : Light nodes update blockchain record along with the newest data
Step 11 : End

8.5 Another dimension of BIIoT architecture

Figure 8.4 shows another dimension of BIIoT architecture. A blockchain composite layer acts as the middleware among industrial applications and IoT. This plan contains two advantages: 1) providing the abstract from lower layers at IoT; or 2) offering services based on blockchain with users. Significantly, a blockchain composite layer obscures a diversity of the underlying layers (such as the communication layer and the perception layer in IoT). A usual IIoT scheme contains the following layered subsystems:

- **Perception layer:** A wide variety of IoT devices include smart meters, radio-frequency identification (RFID) tags, controllers, actuator and sensors. These devices can sense and gather information from the physical surroundings. In the meantime, a few of them (like controllers and actuators) can influence actions in the surroundings [11].
- **Communication layer:** A variety of wired/wireless devices for example controllers, actuators, RFIDs, sensors, and other tags could then link to IoT gateways macro BS, small base stations (BS) and WiFi Access Points (APs) to create the industrial network. A network link is allowed through the separation of contact protocols, for example, Low-Power Wide-Area Network (LPWAN) technology containing industrial Ethernet, Narrowband-IoT (NB-IoT), LoRa, with Sigfox [12].
- **Industrial applications:** IoT can broadly be utilized to sustain numerous industrial applications. The usual industrial appliances comprise internet of vehicles, health care, smart grid, food industry, supply chain, and manufacturing.

The blockchain-composite layer provides numerous services based on blockchain, application programming interfaces (APIs) to assist different industrial applications [13]. Consequently, the blockchain-composite layer could further decrease the trouble in creating industrial applications because of the abstraction.

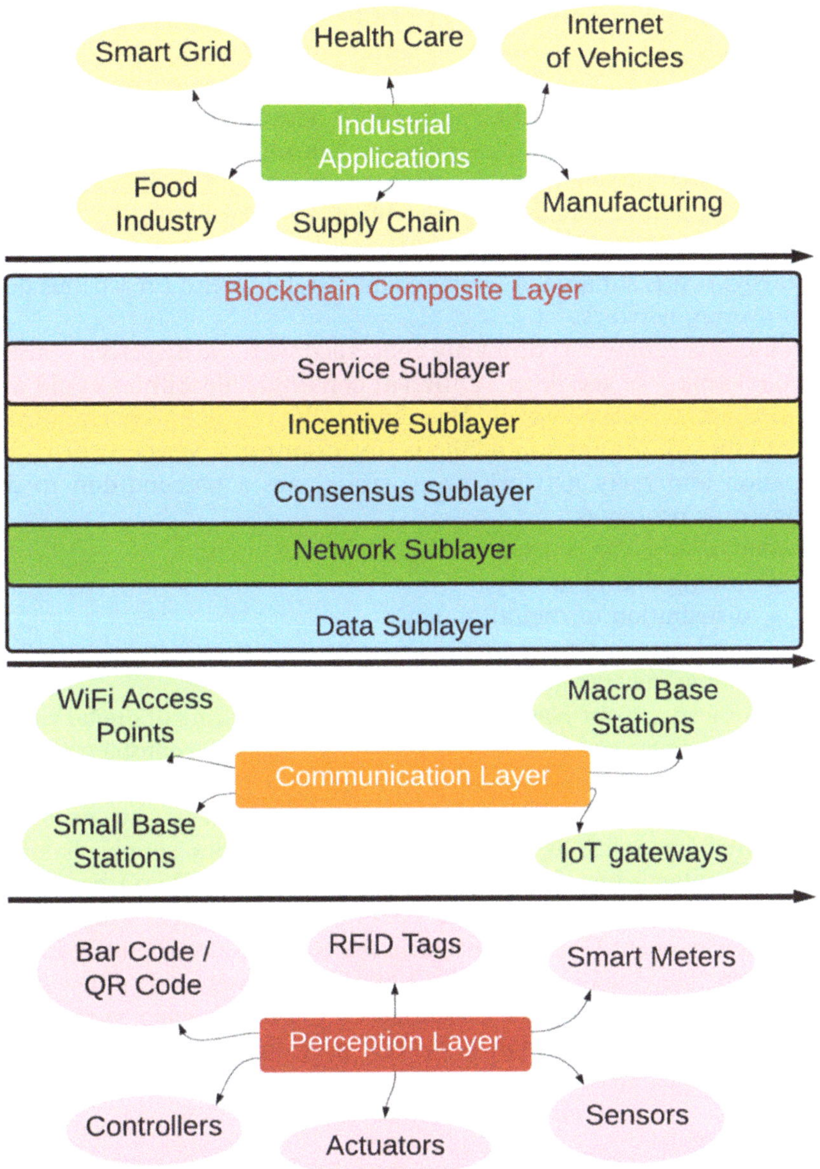

Figure 8.4. Another dimension of BIIoT architecture.

Specifically, the blockchain-composite layer comprises of five sublayers as demonstrated in figure 8.4 (from bottom to top):

(1) The data sublayer gathers IoT data from the lower layers (for example, the perception layer) and encloses the encrypted data with digital signatures through hash functions and asymmetric cryptographic algorithms. These continuously linked data blocks make the blockchain after distributed verification. Various blockchain platforms could select multiple hash

functions and cryptographic algorithms. For instance, the bitcoin blockchain selects ECDSA as the signature algorithm and SHA-256 as a hash function.

(2) A network sublayer is the overlay P2P network executing over a layer of communication. An overlay network comprises either physical or virtual connections linking nodes in the underlying transmission networks. One device basically only transmits a block of transactions for its associated peers. Having received a block of marketing, further peers would validate it nearby. If it is suitable, a block would also be spread for various devices by an overlay network.

(3) Consensus sublayer is primarily concerned with the dispersed consensus for the reliability of the block. Different consensus algorithms could attain the consensus as DPOS, PBFT, PoS, and PoW [14]. It is important to mention that block dissemination methods (for example advertisement-based propagation and relay network propagation) are a precondition to dispersed consensus protocols.

(4) Incentive sublayer is accountable for the next jobs:
- issuing digital money;
- distribution of digital money;
- implementing prize method (particularly to miners);
- managing transaction price and so on.

Mainly, its significance is for planning the suitable financial rule of digital currency (for example, money-making and sharing), sharing rewards to the contributors who participate in spreading consensus (for example, mining).

(5) Service sublayer offers users with services based on blockchain for different industrial areas such as the food industry and utilities, supply chains, logistics and manufacturing. The blockchain as a service (BaaS) can be attained through smart contracts, that can be routinely encouraged while the particular occasion happens [15]. For instance, the fee deals are routinely implemented while the merchandise is obtained by the customer. It is noteworthy that a network sublayer which is recognized on top of a layer of communication is underneath communication networks, providing worldwide network access across various networks.

8.6 Summary

The IIoT updates smart industry through initiating the latest technology; also, blockchain is a growing method which has attracted considerable awareness from across industry, and even academia. Combining blockchain with smart industry attaches numerous superior characteristics, such as automation, decentralization, trust and security. This chapter discussed blockchain with an IIoT architecture (BIIoT) for smart industrial surroundings. The BIIoT architecture allows a safe, lightweight, and decentralized private blockchain-based IIoT network that executes numerous vital processes, for example, a device with user enrollment, information storage, even machine processing in a dependence method. The next chapter will discuss the deployment of BIIoT.

References

[1] Kumar S and Jasuja A 2017 Air quality monitoring system based on IoT using Raspberry Pi *In 2017 Int. Conf. on Computing, Communication and Automation (ICCCA)* (Piscataway, NJ: IEEE) pp 1341–6

[2] Rouzbeh B, Bone G M, Ashby G and Li E 2019 Design, implementation and control of an improved hybrid pneumatic-electric actuator for robot arms *IEEE Access* **7** 14699–713

[3] Liu Y, Wang K, Lin Y and Xu W 2019 LightChain: a lightweight blockchain system for industrial Internet of Things *IEEE Trans. Ind. Inf.* **15** 3571–81

[4] Xinyi Y, Yi Z and He Y 2018 Technical characteristics and model of blockchain *2018 10th Int. Conf. on Communication Software and Networks (ICCSN)* (Piscataway, NJ: IEEE) pp 562–6

[5] Latif S, Idrees Z, Ahmad J, Zheng L and Zou Z 2020 A blockchain-based architecture for secure and trustworthy operations in the IoT *J. Ind. Inf. Integr.* **21** 100190

[6] Sahoo M, Singhar S S, Nayak B and Mohanta B K 2019 A blockchain-based framework secured by ECDSA to curb drug counterfeiting *2019 10th Int. Conf. on Computing, Communication and Networking Technologies (ICCCNT)* (Piscataway, NJ: IEEE) pp 1–6

[7] Simos D E, Mekesis S, Kuhn D R and Kacker R N 2017 Combinatorial coverage measurement of test vectors used in cryptographic algorithm validation *STC'17: Proc. of the 2017 IEEE Software Technology Conf.*

[8] Cao B, Wang X, Zhang W, Song H and Lv Z 2020 A many-objective optimization model of industrial internet of things based on a private blockchain *IEEE Netw.* **34** 78–83

[9] Mingxiao D, Xiaofeng M, Zhe Z, Xiangwei W and Qijun C 2017 A review on consensus algorithm of blockchain *2017 IEEE Int. Conf. on Systems, Man, and Cybernetics (SMC)* (Piscataway, NJ: IEEE) pp 2567–72

[10] Abdi A I, Eassa F E, Jambi K, Almarhabi K and AL-Ghamdi A S 2020 Blockchain platforms and access control classification for IoT systems *Symmetry.* **12** 1663

[11] Tukur Y M, Thakker D and Awan I U 2019 Ethereum blockchain-based solution to insider threats on perception layer of IoT systems *2019 IEEE Global Conf. on the Internet of Things (GCIoT)* (Piscataway, NJ: IEEE) pp 1–6

[12] Kan L, Wei Y, Muhammad A H, Siyuan W, Linchao G and Kai H 2018 A multiple blockchains architecture on inter-blockchain communication *2018 IEEE Int. Conf. on Software Quality, Reliability and Security Companion (QRS-C)* (Piscataway, NJ: IEEE) pp 139–45

[13] Xu X 2020 Towards automated migration for blockchain-based decentralized application *In 2020 IEEE/ACM 42nd Int. Conf. on Software Engineering: Companion Proc. (ICSE-Companion)* (Piscataway, NJ: IEEE) pp 155–7

[14] Wan S, Li M, Liu G and Wang C 2019 Recent advances in consensus protocols for blockchain: a survey *Wirel. Netw.* **26** 5579–93

[15] Ochoa J J, Bere G, Aenugu I R, Kim T and Choo K K 2020 Blockchain-as-a-Service (BaaS) for battery energy storage systems *2020 IEEE Texas Power and Energy Conf. (TPEC)* (Piscataway, NJ: IEEE) pp 1–6

IOP Publishing

Lakshmana Kumar Ramasamy and Seifedine Kadry

Chapter 9

BIIOT deployment

IIoT systems face many challenges, for example, security and privacy vulnerabilities, and resource restrictions on IIoT devices, low mobility, and diversity of IIoT systems. Blockchain technologies may complement IIoT systems with enhanced functionality and enhanced security and privacy. Furthermore, blockchain may improve the scalability and reliability of IIoT systems. In previous chapters, we discussed the combination of blockchain with IIoT as BIIoT. This chapter discusses the deployment of BIIoT.

9.1 General deployment of BIIoT

The real deployment of BIIoT is significant. However, because of a restriction of IoT devices, it is difficult to record an entire blockchain in IoT devices. Significantly, there are two modes to record the blockchain information: (i) full storage, wherein the whole blockchain is recorded, (ii) partial storage, wherein only a subset of data blocks is recorded locally. Consequently, we call nodes with fully occupied storage of blockchain information full nodes and nodes with partially occupied storage of blockchain information lightweight nodes. The full node could be the edge server or the cloud server accompanied by sufficient calculating resource. This is because it needs vast storage space to store a whole blockchain (for example, the entire bitcoin blockchain engaged almost 185 GB by the end of September 2018) and extreme calculating ability to resolve consensus puzzles (for example mining). Conversely, assets restrained IoT nodes (for example sensors) could be lightweight devices that could authenticate the transaction's reliability in the absence of storing or downloading an entire blockchain (for example, merely storing partial blockchain information, for example hash value). It is noteworthy that lightweight nodes depend more on full nodes. Figure 9.1 provides the feasible deployment situation of BIIoT, at which edge servers with cloud servers might store all blockchain (otherwise partial blockchain) information simultaneously while IoT devices might merely store partial blockchain information.

doi:10.1088/978-0-7503-3663-5ch9

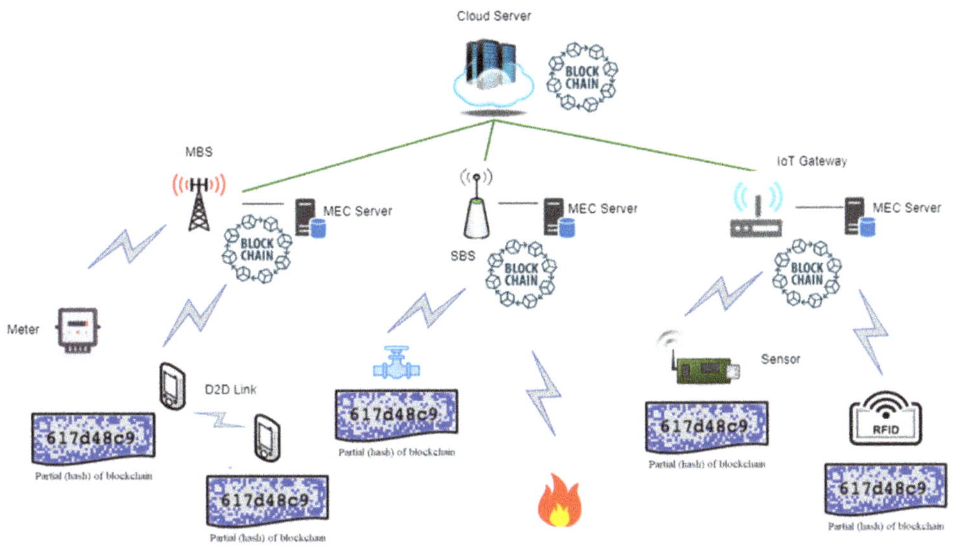

Figure 9.1. BIIoT deployment situation.

There are numerous feasible communication methods among blockchain and IoT as well as BIIoT deployment:

(i) direct communication among blockchain and IoT, wherein IoT devices could directly use blockchain information storage in edge servers in co-location with Small BS, Macro Base Stations (MBS) or IoT gateways;

(ii) straight communication among IoT devices, wherein IoT devices could directly use/swap partial blockchain information through D2D connections;

(iii) hybrid communication of edge and cloud servers accompanied by IoT nodes, wherein IoT nodes could communicate by blockchain information via cloud/edge servers.

There are many attempts to tackle a configuration and launch of blockchain on IoT devices or edge servers. For instance, Raspnode2 is the scheme mostly for installing bitcoin and other blockchains in Raspberry Pi microcomputers. EthArmbian3 provides a custom-made Ubuntu Linux image for ARM nodes, each of which could act like the Ethereum device. Regardless of these efforts, many IoT nodes are still lightweight devices because of constrained storage.

Figure 9.2 illustrates a few decentralized industrial activities with production deployment schemes that can advertise a BIIoT platform's performance.

- **Registry of inventory and assets**: BIIoT can be deployed to create appliances for keeping reports of product inventory and assets. These appliances can maintain accounts of asset detection data and movement of possessions by a supply chain removing the requirement for physical paper reports.
- **Supplier identity tracking and reputation**: BIIoT can be deployed for implementing supplier identity and reputation managing appliances that direct different parameters for performance (for example seller ratings, customer

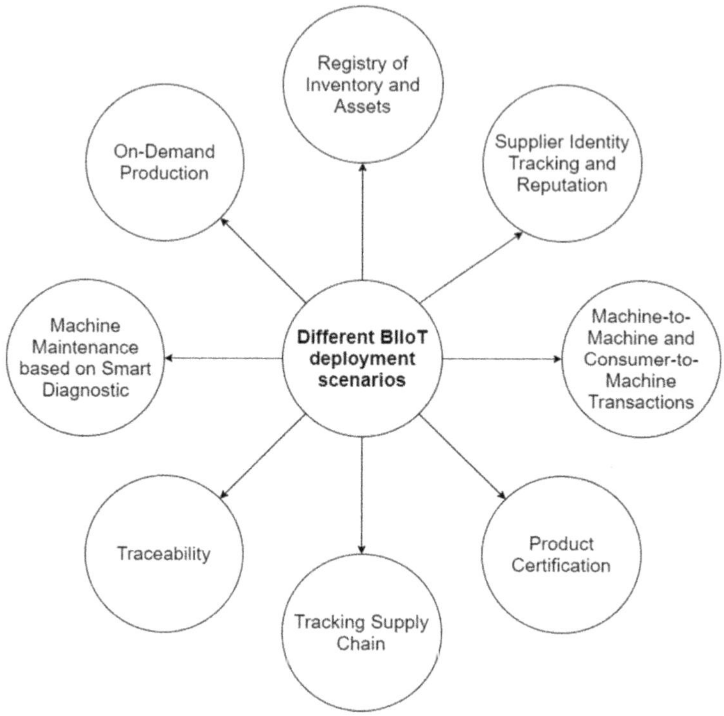

Figure 9.2. Different BIIoT deployment scenarios.

reviews and delivery times) for vendors. A few applications can be utilized by customers to discover vendors who can meet their production needs and by producers to locate retailers of consumer goods. Smart contracts allow you to spontaneously negotiate better costs for consumer goods and products at any time using vendor repute.

- **Machine-to-machine and consumer-to-machine transactions**: BIIoT can allow machine-to-machine dealings for production services. For instance, the customer can transmit the demand for producing the manufactured goods through transferring the transaction to a producer's smart contract and complete their payment with cryptocurrency (for example, Ethers). The producer's smart contract can transmit smart contracts related to individual machines (consumer-to-machine transactions). If various machines' services are needed for producing manufactured goods, devices could transfer micro-transactions to multiple machines (machine-to-machine transactions).
- **Product certification**: BIIoT can be deployed for implementing product certification appliances. The product data for manufactured goods (for example, producing ability information, machine information, producing date with parts data) can be registered in a blockchain. This data could help verify the genuineness of goods, removing a necessity for physical certificates that could be prone to falsification and interference.

- **Tracking supply chain**: BIIoT can be deployed for implementing tracking supply chain appliances. A blockchain with smart contracts can maintain the proper registry of goods and trace their ownership at various points in the supply chain. A few applications can allow automatic monetary settlement in delivery verifications.
- **Traceability**: BIIoT can be deployed for implementing traceability appliances for developed goods. Smart contracts among producers and customers can maintain manufacture reports, for instance, which machines and which plant inside a factory is utilized for producing specific manufactured goods. Recall of manufactured goods is also controlled (either because of fabrication shortcomings or functional elements defects) after goods are distributed, and traceability appliances can assist in revealing distressed products.
- **Machine maintenance based on smart diagnostics**: BIIoT can be deployed for implementing machine maintenance based on smart diagnostics appliances for machines where machines are capable of supervising their condition, detecting troubles, and booking service, filling of consumables, or element substitute requirements for a machine safeguarding seller. Smart contracts among sellers and producers for purchase of goods and service of equipment could help with mechanizing the self-service jobs.
- **On-demand production**: BIIoT facilitates the marketplace of developed services where machines have their own blockchain account where consumers are capable of purchasing and handling goods with devices directly to benefit expanded services at the CBM in an on-demand form. BIIoT can allow peer-to-peer networked developed appliances where peers can benefit from producing services (for example, 3D printing or CNC machining) via transmitting transactions for machines.

Ethereum is a famous program for BIIoT deployments. Ethereum offers more properties than bitcoin, adding smart contracts to enlarge potential appliances considerably.

AeroToken [1] aims to make possible real-time manoeuvring over property with an access approval scheme for low altitude profitable drone services. Poperty owners provide the outcome for drone flying by willingly distributing airspace above their property. It assists in resolving the difficulty of drone function authorization, constructing a novel distribution market. The property owner presents their airspace based on a smart contract in a blockchain, and drone services suppliers pay for provisional usage. Appliances have been implemented based on Ethereum smart contracts.

Aigang [2] is an independent insurance network for IoT possessions. Aigang has organized smart contracts over an Ethereum Testbed which distributes policy, performs risk appraisal and develops requests mechanically. They further use traditional virtual money (AIX). They present investment opportunities in various goods with different risk levels, even possible profits. A smart contract connects a smart device with an insurance policy. By this contract devices can direct preservation and the insurance fee can be computerized. By insertion of Oracles for account proceedings, request managing can be electronically managed.

Chronicled [3] have implemented numerous IoT manufacturing activities that cryptographic ability makes possible with the aim of making the world more dependent on IoT with supply chains environments. Chronicled contains the blockchain management server coordinating by numerous blockchain schemes, for example, Hyperledger, Ethereum with Quorum. Confirmation with the register of device individuality is implemented by a smart contract.

At LO3 Energy [4] the power microgrid which utilizes blockchain has been demonstrated in South Australia and southern Germany. A microgrid is a localizing group of energy production, power storing, with electrical shipments. A scheme constructs the public power market, which makes the decentralized peer-to-peer energy network which is synchronized by a vast energy network. It is a primary power-blockchain platform which permits devices in a network border for safety with an open transaction to power vending between microgrids contributors. The mixed devices calculate the construction's power manufacture in use and transmits information for a grid.

MyBit [5] is designed to construct the environment of service where IoT resources (from cars to drones) are are owned by the community, and income is distributed. This novel finance replica offers savings opportunites open for everyone. An Ethereum smart contract is utilized for mechanization to develop: while an IoT device creates income a depositor automatically obtains the part of earnings proportional to their possession of stock. The middle smart contract is accountable to a directive, maintained with the introduction of a proposal. This proposal describes various resource kinds; also IoT devices are connected for resources, and once established they transmit and demand data via the API. Oracles are utilized for linking devices for a network.

A blockchain structure is described through Slock. It [6] aims to tackle confidentiality, synchronization, identity with safety of over a billion IoT devices. An objective is to construct the distributing wealth where every IoT resource could be borrowed safely by rapidly connecting to some power. They are functioning in the diversity of schemes, for example, the market for the present charge communications to an electric vehicle, named Block charge. This result utilizes the smart plug, the mobile app to start a plug with managing an account, also a blockchain for disbursing to a service. They are further functioning in the intelligent lock that automates the members' fee.

A Blockchain of Things [7] gives a safe free contact podium to industrial IoT incorporation. They suggest Catenis, the web services layer for quick bitcoin blockchain incorporation, by end encryption. Furthermore, it could further be accommodated for Hyperledger, Ethereum, and various other blockchains. At Catenis, every IoT node is symbolized as a virtual device at the Catenis Hub and Gateways (with increased safety). Every virtual machine would run the host of Catenis service to an IoT device.

A Chain of Things [8] is the blockchain-permitting IoT study lab which suggests Maru, the incorporated IoT hardware with blockchain resolution. Blockchain has always offered devices worldwide individuality, safety with functionality. Three case studies have shown this: Chain of Protection, concentrated in presenting security for

IoT via blockchain; Chain of Solar, which aims to link solar panels for a blockchain for saving generated power in a diversity of appliances; and Chain of Shipping, which is designed for enhancing security of delivery with the logistics industry. Data logging devices are utilized to transmit information for a network. The proof of capacity (PoC) has been implemented over an Ethereum network.

Modum [9] blockchain aims to provide information reliability to physical products with a focus on improving supply chain processing. A Modum sensor registers an ecological situation through consignments. Modum is designed to work with various platforms. The resolutions to a sharing of medical products have implemented based on Ethereum blockchain. An Ethereum smart contract is utilized to confirm sensor information every time merchandise changes ownership, a contract authenticates which transaction meets a consumer's values. A result combines a sensor tag which gathers quantifications during transit, the mobile app, utilized for a link with a smart sensor and change of ownership. Moreover, the control panel for examining the sensor collects information after the delivery confirmation.

A scheme called Autonomous Decentralized P2P Telemetry (ADEPT) led by Samsung with IBM [10] aims at encouraging device independence; moreover, they utilize blockchain to guarantee code implementation in the edge device. ADEPT uses three protocols: Ethereum, Telehash, and BitTorrent, for blockchain, messaging and file distribution, respectively. Blockchain presents checklists, contracts, engagement and authentication. Their PoC comprises an intelligent washing machine that utilizes a smart contract for purchasing detergent provision from vendors.

Twin of Things is the result of protecting a possession with the origin of daily objects [11] implemented through Riddle and Code. The work merges cryptography with blockchain to create the digital identity based on hardware to an entire linked physical thing. Contact with dealings among devices is independent with safely executed as a result of blockchain. The extremely safe crypto chip allows every device to become a blockchain device. A chip created in the Glue inevitable near field communication (NFC) tag with an Android app is used to exchange blockchain for registration for a unique, undamaged ID for a chip. Once authenticated in a network, it can communicate with a different device. The obtained results can presently work with BigchainDB blockchain Ethereum and Bitcoin.

9.2 Deployment of BIIoT in smart industry

This section explains the steps of initialization, handling transactions, and shared overlay.

9.2.1 Initialization

This section discusses the implementation of adding a policy header and devices with a local blockchain. To attach the device for smart industry, a miner creates the genesis transaction through distributing the key with a device utilizing the widespread Diffie–Hellman key exchange. A distributed key among miners and an

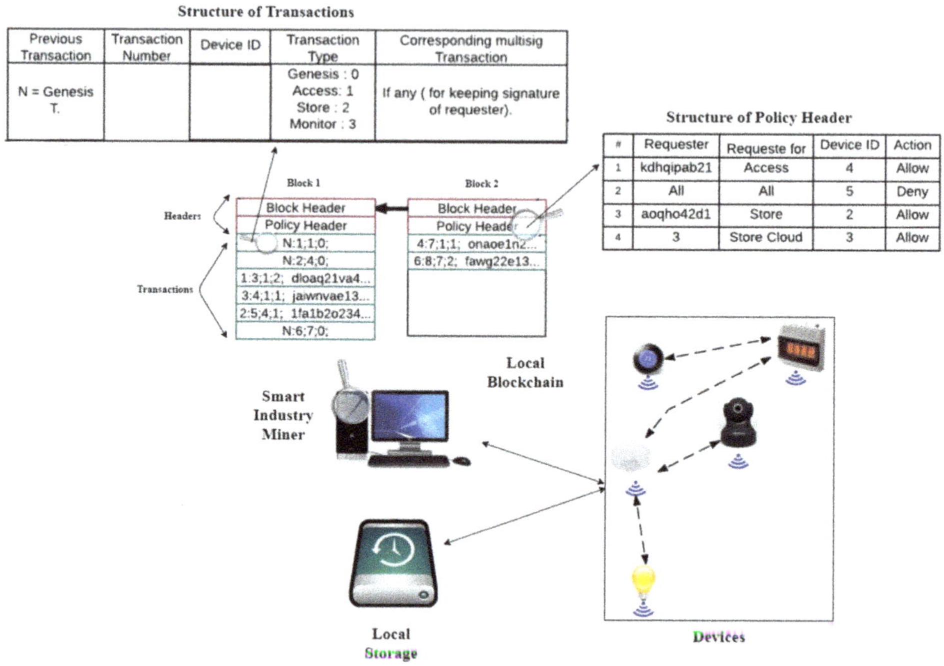

Figure 9.3. Deployment of BIIoT in smart industry.

instrument is recorded in a genesis transaction. To explain the policy header, an industry proprietor creates their own policy and proposed policy structure, as in figure 9.3, and attaches a policy header for a first block. A miner utilizes a policy header at the newest block in the blockchain; thus, to adapt a policy, a possessor must update a new block's policy header.

9.2.2 Handling transactions

Smart devices might be contacted directly by everyone or by an entity outside for an intelligent industry. Every device within an enterprise might demand information from another internal device for specific services. For example, a light bulb requires data from a movement sensor to switch on lights mechanically when somebody enters an building. To attain consumer control over smart industry dealings, the shared key must be allotted by a miner for a device that requires direct contact with everyone.

To allot a key, a miner verifies a policy header or requests consent from an possessor and subsequently issues the shared key among devices. Once getting a key, devices communicate directly providing their key is legitimate. To reject consent, a miner marks a shared key as invalid by transmitting the control message for devices. Benefits for this technique are two-fold: first, a miner (thus a possessor) has the devices list which distributes information; and second, communication beween devices is secure with the shared key.

Recording information in local storage with devices is a different feasible flow of transaction within an industry. To record information locally, every device wants to be valid for storage which is completed utilizing the shared key. To allow a key, a device needs to transmit the request to a miner and if it has recording consent, a miner creates the shared key and transmits a key for a device with storage. Through getting a key, neighbouring storage now has a shared key. Having a shared key allows a device to record information directly to local storage.

Devices might require storing information in cloud storage which is recognized as storing transactions. Recording information in a cloud is an anonymous procedure. To record information, a requester wants the origin which has the block-number with the hash utilized for an unknown verification reason. Cloud storage might also be possessed directed through a service provider (SP) (for example Nest thermostat) or rewarded for and required through an industry possessor (for example Dropbox).

In the previous example, the miner demands an origin via creating the signed transaction by a device key. In this case, payment is completed via bitcoin. In any storage kind, after getting the request for storage it generates the origin and transmits it for a miner. While the device requires recording information in cloud storage, it communicates a report and an instrruction for a miner. Through getting an order, miners authorize a device to record information in cloud storage. If a device has approved, a miner takes out a final block-number and hash from a local blockchain, and generates the record transaction and transmits it in addition to information for storage. After recording information, cloud storage returns a novel block number for a miner which is utilized for more recording transactions.

A further feasible transaction is usage with supervising transactions. These transactions are mostly created by either an industry possessor for overseeing an industry while he is external or through SPs to develop devices' information to adapted services. Through getting the usage transaction from devices in an overlay, a miner verifies whether the requested information is in local or cloud storage. If the information is recorded in local storage, a miner needs input from local storage and transmits it for a requester. Conversely, if the data is recorded in a cloud, a miner either demands information from cloud storage and sends it for a requester or transmits a final block number with a hash for a demander.

A final situation authorizes a demander to study complete information recorded through a device in cloud storage and is appropriate while recorded data is to the specific device. Otherwise, a customer's confidentiality may be in danger as a fraction of the connecting attack. Through getting the monitor transaction, a miner transmits present information of a demanded node for a requester. If the requester is permitted to obtain information at that time then a miner transmits information until the requester transmits the close demand to a miner thereby eliminating a transaction. A monitor transaction allows a possessor to monitor cameras or various devices that transfer intervallic information. To evade reasonable attacks, a possessor must describe the threshold in minutes for intervallic information. When the time comes for the miner to send the data to the claimant, the connection will be stopped by the miner.

9.2.3 Shared overlay

While individuals operate over more than one industry, they require different miners and storage to every one of the sectors. To reduce cost of administration overheads in this example, the shared overlay is described. A shared overlay contains a minimum of two smart industries directed centrally like a solitary industry through a shared miner. A shared overlay is related to an intelligent industry; however, an organization of a distributed blockchain is dissimilar to smart industry.

In a distributed blockchain, every industry has the genesis transaction, and a genesis transaction of all devices is chained for their industries genesis transaction through a shared overlay miner. One more difference in a shared overlay concerns contacts among industries by a miner. Appliances in a similar sector by miner knowledge do not change, whilst for devices in other sectors, a Virtual Private Network (VPN) link is recognized among an Internet gateway at every industry with a miner of a shared overlay which routes packets for a distributed miner.

9.3 Key challenges of BIIoT

While blockchain may seem capable of the IIoT, there are numerous challenges to ensuring its well-informed acceptance. The main challenges are as follows, and shown in figure 9.4.

- **Awareness**: Blockchain is a new technology widely accepted in the monetary division (bitcoin is the famous appliance). Lack of awareness regarding blockchain in other domains affects its acceptance.
- **CAP and blockchain**: For distributed data systems, there is an exchange between stability and availability. These trade-offs are illustrated with the consistency, availability and partition tolerance (CAP) theorem, which states that under a partition, the distributed data system may be uniform or available, but not both at the same time. Blockchain provides general stability and shared tolerance. Blockchain is a distributed ledger that is equally identical, meaning that all nodes eventually find the same ledger. In Ethereum, the block time (a new block cut time) is approximately 17 s, 10 min faster than bitcoin block time. Security is reduced due to speed block time; therefore, many blockchain applications require multiple confirmations for newly mined modules to protect transactions from double spending. Quick block time leads to the production of a large number of older blocks. Older blocks are competing blocks made by miners that do not contribute to the main chain. The more primitive rate reduces the security of the main chain.
- **Competence**: This is not an efficient approach as all nodes in the blockchain network perform the same calculations to mine the block next to the blockchain. Due to this redundancy in operation, the contribution of an individual node to the overall network is minimal, even if the node performs complex calculations.
- **Confidentiality**: Because blockchain is a public ledger, and anyone can view all transactions within the blockchain network, there are privacy concerns for the parties to the transaction.

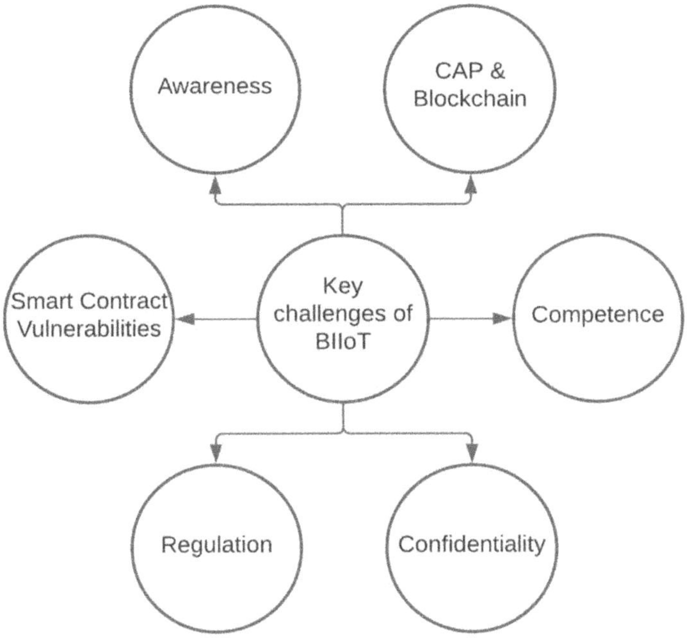

Figure 9.4. Key challenges of BIIoT.

- **Regulation**: Since blockchain eliminates the need for a centralized authority or reliable intermediary to verify transactions, there is regulatory urgency in the widespread adoption of the technology. Decentralized organizations such as blockchain need new government and industry regulations. Furthermore, there is a need to ensure the legal implementation of smart contracts to avoid disputes between the parties to the transaction.
- **Smart Contract Vulnerabilities**: Smart contracts can have software vulnerabilities that can be exploited by hackers. In June 2016, an attacker was able to expel more than 3.6 million ethers from Slock.it, supporting the decentralized autonomous organization (DAO) as a 'child DAO' [12]. The attacker managed this by using the 'recursive call error' vulnerability in the DAO smart contract. Since smart contracts are contracts between parties that trade on a blockchain and cannot legally be enforced outside the network, such attacks could put companies, miners and even the blockchain network at risk.

9.4 Summary

BIIoT can securely exchange and store data from components in IIoT systems without the need for an intermediary. The security of blockchain has enabled IIoT systems to run more reliably and securely, which changed the way data was handled in IIoT. This chapter discussed the general deployment of BIIoT and deployment of BIIoT in smart industry. The next chapter will discuss the BIIoT applications in summary.

References

[1] Perkins C D, Wroble E J and Halpin H A 2017 Creating a drone superhighway using the blockchain https://neironix.io/documents/whitepaper/7764683f65f9877790bb6c3350b3cf87.pdf
[2] Chandel S, Zhang S and Wu H 2020 Using blockchain in IoT: is it a smooth road ahead for real? *Future of Information and Communication Conference* (Cham: Springer) pp 159–71
[3] Ullah H S, Aslam S and Anrjomand N 2020 Blockchain in healthcare and medicine: a contemporary research of applications, challenges, and future perspectives (arXiv preprint arXiv:2004.06795)
[4] Park I, Lee T and Jang J 2020 Trade-off in implementation of P2P energy trading over hyperledger blockchain *2020 IEEE Int. Conf. on Consumer Electronics-Asia (ICCE-Asia)* (Piscataway, NJ: IEEE) pp 1–4
[5] Chitchyan R and Murkin J 2018 Review of blockchain technology and its expectations: case of the energy sector (arXiv preprint arXiv:1803.03567)
[6] Karafiloski E and Mishev A 2017 Blockchain solutions for big data challenges: a literature review *IEEE EUROCON 2017—17th Int. Conference on Smart Technologies* (Piscataway, NJ: IEEE) pp 763–8
[7] Abadi F A, Ellul J and Azzopardi G 2018 The blockchain of things, beyond bitcoin: a systematic review *2018 IEEE International Conference on the Internet of Things (iThings) and IEEE Green Computing and Communications (GreenCom) and IEEE Cyber, Physical and Social Computing (CPSCom) and IEEE Smart Data (SmartData)* (Piscataway, NJ: IEEE) pp 1666–72
[8] Chain of things, available online: https://chainofthings.com/ (accessed 21 January 2021)
[9] Katuwal G J, Pandey S, Hennessey M and Lamichhane B 2018 Applications of blockchain in healthcare: current landscape & challenges (arXiv preprint arXiv:1812.02776)
[10] Veena P, Panikkar S, Nair S and Brody P 2015 Empowering the edge-practical insights on a decentralized internet of things, empowering the edge-practical insights on a decentralized Internet of Things *IBM Institute for Business Value* **17**
[11] Riddle and code available online: https://riddleandcode.com/poc (accessed 21 January 2021).
[12] DuPont Q 2017 Experiments in algorithmic governance: a history and ethnography of 'The DAO,' a failed decentralized autonomous organization *Bitcoin and Beyond* ed M Campbell-Verduyn (London: Taylor & Francis) pp 157–77

Chapter 10

BIIOT applications

There is a rising trend for blockchain in IoT because blockchain can help conquer its challenges. This chapter provides the general idea of BIIoT applications. It is important to note that there is a broad variety of blockchain applications (ranging from smart production to Internet of Vehicles (IoV) with unmanned aerial vehicles (UAV)). This chapter focuses primarily on BIIoT industrial applications. Figure 10.1 shows seven main applications of BIIoT.

10.1 Smart manufacturing

Manufacturing businesses are experiencing advancement from automatic manufacturing to 'smart manufacturing' [1]. Big data analysis of production information acts as a significant part of this upgrade development. An enormous amount of information is created at each stage of the product life cycle, including design of the product, delivery of components, production, delivery, and trade, with service after-sales. Although the production data is very piecemeal, it is essential to data analysis complexity with data aggregation. BIIoT can tackle the functionality problem by interconnecting IoT through the peer-to-peer network thereby permitting data distribution across the industrial sector. For instance, widely disseminated blockchain can be built to provide for various sectors with each blockchain providing for an industry or more than one sector. BIIoT can further enhance the safety of smart manufacturing. One of the main impediments restraining the improvement of the factory is that the IoT has to be maintained by a centralized method. For instance, IoT firmware requires being alerted frequently to a solution safety violation. However, many of the firmware upgrades are downloaded from the central server and subsequently physically installed on the IoT device. It is costly and inefficient to install and upgrade the firmware updates in disseminated IoT. The technique of Kusiak [2] provides an automated firmware improvement result using blockchain and smart contracts. Significantly, intelligent contracts explain the firmware improvement ways (for example, where and when to improve firmware)

doi:10.1088/978-0-7503-3663-5ch10

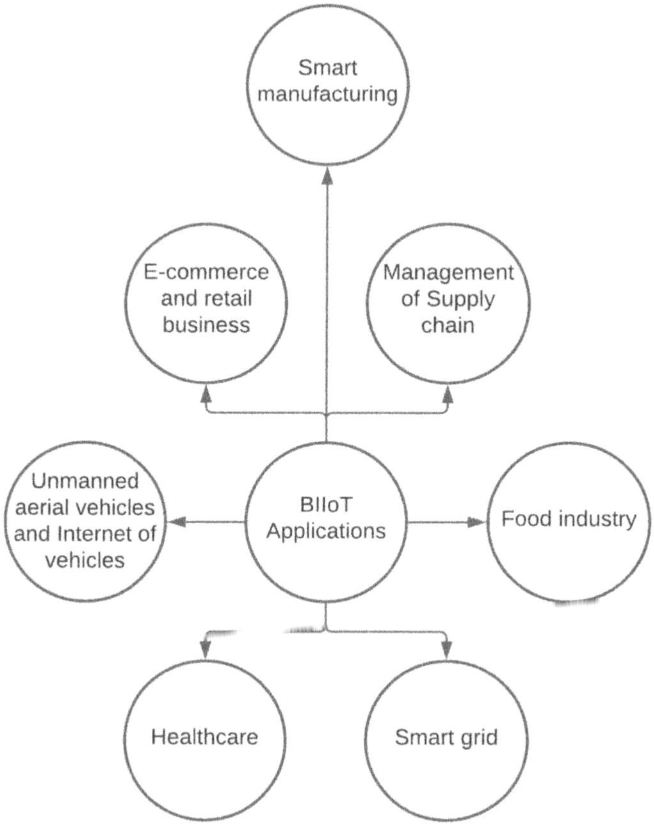

Figure 10.1. Applications of BIIoT.

positioned over the entire industrial network. Devices can then download and install the firmware hash through the smart contract to be implemented spontaneously. Consequently, the safety preservation fee can be very much lower.

Additionally, Wan *et al* [3] presented a decentralized blockchain-based automated manufacture proposal for superior safety with confidentiality defence over traditional centralized architecture. A mobile crowdsensing scheme using blockchain was introduced to resolve the incentive problem by an information excellence guarantee in smart manufacturing [4].

10.2 Management of supply chain

Manufactured goods frequently consist of numerous parts presented by various manufacturers across nations. However, a few low-quality parts might seep into the supply chain. The use of anti-fraud technology in every piece of manufactured goods is very costly. The combination of blockchain with IoT could resolve this difficulty. Significantly, each part would be labeled with a unique ID from formation. Meanwhile, an unchangeable time-stamp is further attached with this ID. The discovery of each piece can be stored within a blockchain that is detectable and tamper-resistant. For instance, the technique of Konstantinidis *et al* [5] demonstrates that the part possession of manufactured goods can be authenticated via a system based on blockchain.

Furthermore, the technique of Kim and Laskowski [6] provides provenance ontology by a combination of IoT with blockchain using Ethereum blockchain. BIIoT structures have shown for assurance the information origin of the supply chain. Conversely, BIIoT can be utilized to decrease fees in post-sales services in supply chain management. The work of Tapscott and Tapscott [7] illustrates the consumer case of motor insurance where claims solution can be automatic by smart contract using blockchain, thus enhancing performance and decreasing claim execution time. Furthermore, it is explored by Kusiak [2], how combining IoT and blockchain can assist in reducing the fee, improving speed and decreasing the danger in the management of the supply chain. Moreover, a machine learning platform using blockchain [1] was presented to secure information distribution between various enterprises to enhance consumer service excellence.

10.3 Food industry

BIIoT can improve the viewable condition of manufactured goods' life cycle, particularly in the food industry. Notably, food products traceability is essential for food security. There is demand for a current IoT for assurance in food traceability throughout the food supply chain [3]. For instance, the strength of the food industry is an arrangement with numerous providers. Traceability needs digitized data of ingredients from the source to each department of food production. During this process, blockchain can confirm traceability and the origin of food business information. There are numerous schemes in this feature. For instance, the work of Tian [8] presented the utilization of radio-frequency identification (RFID) as well as blockchain for setting up the supply chain from agriculture to food manufacturers in China. It demonstrated proof of traceability of food supply chain information.

The work of Sander *et al* [9] demonstrated a blockchain can assist in enhancing food security through the stipulation of the detectable food items. Furthermore, it is revealed by Bettín-Díaz *et al* [10] that a blockchain combination in food supply chain can permit clients to trace an entire food manufacturing procedure. They further provide the consumer case of utilizing blockchain for natural coffee production in Colombia. Moreover, Lin *et al* [11] present a food security traceability scheme using a blockchain with IoT tags of Electronic Product Code (EPC). Significantly, it can avoid information tampering with an expression of privacy by the smart contract. The model of a presented structure is developed to show efficiency.

10.4 Smart grid

The origin of disseminated renewable energy assets is redesigning energy customers' role from pure customers to prosumers who can generate energy (for example, from renewable energy assets) and consume it [12]. Energy suppliers with extra energy can sell it to other customers. We call this trade between consumers and prosumers (for example peers) peer-to-peer power trade. However, it is demanding to ensure security and trust of a power trade between two trading parties in a shared environment. The emergence of blockchain brings opportunities to ensure safe peer-to-peer power trade. Several studies have been presented utilizing blockchain to address these challenges. For instance, the work of Li *et al* [13] implemented a safe power trade scheme using a consortium blockchain. This scheme could secure the trade price without going via a central broker through the disseminated consensus of blockchain.

Furthermore, Aitzhan and Svetinovic [14] implemented a decentralized power-trade scheme using blockchain. This scheme showed the efficiency of securing secret energy-trading transactions in decentralized smart grid schemes. Moreover, Pop *et al* [15] presented a blockchain-based method to give safe and clear energy demand-side running on the smart grid.

10.5 Health care

Health care is becoming one of the main socioeconomic troubles due to an ageing population. Also, it poses a novel challenge in conventional health service due to limited clinical assets. A current advance in easy-to-wear healthcare devices and big data analytics in healthcare information can promote distant healthcare services at home or in a hospital setting. Consequently, the workload of hospital assets can be alleviated [16]. For instance, senior citizens in their homes are wearing healthcare devices on their body. These wearable devices continuously calculate healthcare information such as blood pressure reading, heart rate and blood sugar. Healthcare teams and doctors can use healthcare information at any time or place through the healthcare network. However, assessing healthcare information brings confidentiality issues, also a safety concern. The susceptibility of healthcare devices and the heterogeneity of the healthcare network poses the challenge to preserve confidentiality and ensure security of healthcare information. Integrating blockchain into healthcare network could conquer the challenge of confidentiality and guarantee safeguarding of healthcare information. For instance, the work of Esposito *et al* [17] demonstrates that utilizing blockchain can protect healthcare information recorded in cloud servers. Also, Griggs *et al* [18] have developed a scheme using blockchain to guarantee secret healthcare information supervision. The healthcare information created through medical sensors can be spontaneously gathered and broadcasted to the plan by implementing a smart contract, as a result sustaining the actual-time patient monitor. During the entire process, confidentiality can be protected under a blockchain.

The work of Zaman *et al* [19] presented a solution based on blockchain to handle personal healthcare information and sustain information distribution betwwen patients, insurance companies, medical centres and hospitals. During the entire procedure, the confidentiality and safety of healthcare information can be guaranteed. Sun *et al* [20] place an attribute-based signature system in decentralized healthcare blockchain schemes. Also, this system can confirm the reliability of healthcare information detection of the healthcare data owner. Conversely, this system can further protect the confidentiality of the healthcare data owner. Work by Rahman *et al* [21] presents an in-home treatment management structure combining IoT with a mobile edge computing system based on blockchain to offer confidentiality with secrecy guaranteed. The experimental outcomes on the prototype show the efficiency of this proposed scheme. Additionally, blockchain further provides traceability for patients who are infected by viruses, for example, Middle East Respiratory Syndrome (MERS), Severe Acute Respiratory Syndrome (SARS) and COVID-19 [22]. Affected patients wearing IoT devices can be traced at their trajectories; therefore, responses such as self-isolation can be performed, whilst patient confidentiality can be protected through blockchain.

In healthcare, vital patient information can be effectively shared utilizing blockchain, which improves the deliverance of health services, for instance, by decreasing errors in patient care and decreasing the possibility for mismatched patients [23]. It has also been shown that interaction efficiency can be improved using blockchain [24]. The latter can present real and secure access to stakeholders involved in patient health records. Furthermore, blockchain-enabled IoT has been developed to aid supervising and managing diseases. Wearable devices (for monitoring key features and presenting comments), and smart tablets with advanced quality control are a few of its applications. They facilitate checking efficacy and side effects, and monitor medical medications during their trial, exclusive of the danger of customized outcomes [25].

The primary research efforts of blockchain in the field of health are as follows:

A. **Management of patient information**

One of the main challenges encountered in keeping patient reports is where a patient's physical condition associated information is scattered over different health providers. A plan of electronic health record (EHR) was never intended to run lifespan health reports broadening across other organizational systems. The challenges in interoperability obtainable among different service providers and hospitals make an extra obstacle for the successful distribution of information. This lack of integrated information supervision and information transfer indicates how these health reports are not incorporated, however structured. Blockchain helps in presenting the structure for information distribution alongside the safety of this information [26]. In this method, healthcare providers can collect information from a patient which might contain their name, preceding procedures executed, and prescriptions followed and so on. This information is recorded in cloud computing schemes or in the association's available database. The cryptographic hash of information from every information source is calculated and transmits through the public ID of patients for the record in a blockchain. Smart contracts are helpful in this case to run right of entry for patient information. By the API, stakeholders of healthcare can search a blockchain and locate this information without disclosing the ID of a patient. The patient can distribute his/her total medical record with any stakeholder if needed [27].

Another study presented by a researcher in this domain [27] obtains information origin and open information distribution between all participating parties. They offer a scheme named 'MeDShare' that uses a blockchain network accompanied by smart contracts and side-blocks. Side-blocks have smart contract records and assist in keeping the precision of the logs, and preserving block effectiveness. Two algorithms presented by Wu and Tsai [24] use a blockchain scheme to secure the confidentiality of distributed information in the health industry. The primary method utilizes a public key cryptography method for each device presumed to contain the private–public key pair. The method that follows utilizes bilinear key pairs using elliptic-curve-cryptography (ECC) to attain confidentiality with shared management of key. A safe scheme based on blockchain for pervasive social networks (PSN) was presented in [28]. Rather than utilizing direct health information, sensor nodes' addresses with mobile devices are recorded in the blockchain. Also, the approved PSN devices can use physical condition information from this device in a safe way.

B. **Drug traceability**

One of the main problems presently encountered by the pharmaceutical industry is drug counterfeiting, with approximately 10 to 30 per cent of medicines sold across the world not being original. A fake drug market adds up to the sum of 200 billion dollars yearly; however, online purchase of these drugs is reported to be a business worth 75 billion dollars [29]. The main problem of these fake drugs is not knowing these drugs are counterfeit; therefore, they could work in a significantly different way from a primarily predictable drug. It can be dangerous to take fake drugs as inefficient care for the disease is likely to occur. Two main problems of drug traceability can be handled utilizing blockchain. Initially, the pharmaceutical company can supervise their drugs in this supply chain; simply discovering every counterfeit drug in the chain. Next, the stakeholders, mainly the labs, are approved to utilize their knowledge in every problem by tracking the precise location of medicine. Tamper resistance presented through a blockchain is helpful in attaining drugs' traceability [30]. Each transaction going to the block of a blockchain is time-stamped unchangeably, which makes it suitable for tracing every product and guaranteeing that this data is not changed. For ensuring the authenticity and traceability of the drugs, the industries that are registering manufactured goods in the blockchain need to be dependable. Therefore, a merely private blockchain under the management of a central entity is rational for guaranteeing that these fake drugs are not registered. Thus, the 'drug blockchain' approach of the companies is evidence that a medicine produced through the industry is legitimate and not the counterfeit manufactured goods. Pharmaceutical industries are set up in such a way that actors of a supply chain could perform as miners. They can be distributors, producers, or retailers, as determined by the industry. Every individual who has various privileges depending upon their position in a supply chain lab can record novel medicines, but merchants can merely confirm a transaction. Once the drugs are produced, the hash is made which will have the appropriate information concerning the product. Whenever a medicine moves from one entity to another (for example, from the producer to the distributor), information is obtained recorded at the blockchain, making it easy to trace the medicine. Bocek *et al* [31] demonstrate an advantage of utilizing blockchain with a smart contract in a drug supply chain. Avoiding human interference, a blockchain presented through Modum.io AG operates intelligent contracts to evaluate the humidity; and heat level is detected by IoT sensors through a medicine transport procedure.

C. **Commercial implementations**

The list below discusses well-regarded commercial blockchain executions in the health area:

i. **MedRec**: This [32] is the famous instantaneous healthcare blockchain design for recording EHR that has been adopted successfully. A healthcare provider, the patient, and every third-party payer or insurer participating in the medical case can attach information for a healthcare report of patients.

ii. **Medicalchain**: Medicalchain [33] is the decentralized platform utilized in the UK to store patient information. It uses blockchain to generate EHR, the spotlight in a consumer's requirements whilst maintaining a single precise version of its report. Medicalchain authorizes its consumers to present healthcare providers and hospitals with right of entry to their health information.

10.6 IoV and UAV

Internet of Vehicles (IoV) fundamentally combines vehicles with a pedestrian network, vehicles with an infrastructure network, vehicles with a roadside network and a vehicle-to-vehicle network. Non-trustworthiness with the decentralization of IoV poses a challenge for secure transaction-execution and message-transmission. Combining blockchain with IoV will overcome the above challenge. For instance, Yang *et al* [34] implemented a platform for reliance management at IoV over the blockchain. Incredibly, the honesty of messages could be validating through PoS/PoW consensus implemented through Roadside Units (RSUs). Furthermore, blockchain can be utilized to collectively protect the power and data communications among hybrid electric vehicles in smart grids [35, 36] and electric vehicles [37]. In the future, resource allocation in IoVs could also be enhanced by combining blockchain, mobile edge computing and artificial intelligence [38]. Currently, a UAV interaction network can recompense inadequate coverage of a wireless interaction network [39]. At the same time, UAV can further be utilized to send products [40] and obtain real-time traffic flow information [41]. Furthermore, the current work of Cheng *et al* [42] further demonstrates that UAV can sustain mobile edge computing and content-centric networking. It is difficult to guarantee the reliability in decentralized a non-reliance UAV-network and limit any misconduct [43]. The combination of blockchain and UAV network can ensure mutual trust among UAVs. The study of Kapitonov *et al* [44] implemented an independent platform using Ethereum blockchain to offer reliance management of UAV. Furthermore, IBM [45] currently use copyright to implement a scheme using blockchain to protect confidentiality and guarantee the safety of UAV information. Mainly, blockchain blocks would record the data associated with UAV with the type of model, producer, and restricted areas. As a result, any UAV misconduct discovered can be recognized at the time.

10.7 Retail and E-commerce industry

The retail and E-commerce industry might very soon harvest the profits of advances in blockchain. It is another industry that researchers have begun to explore to verify the compatibility of blockchain. Figure 10.2 illustrates a scenario to utilize blockchain in a retail business.

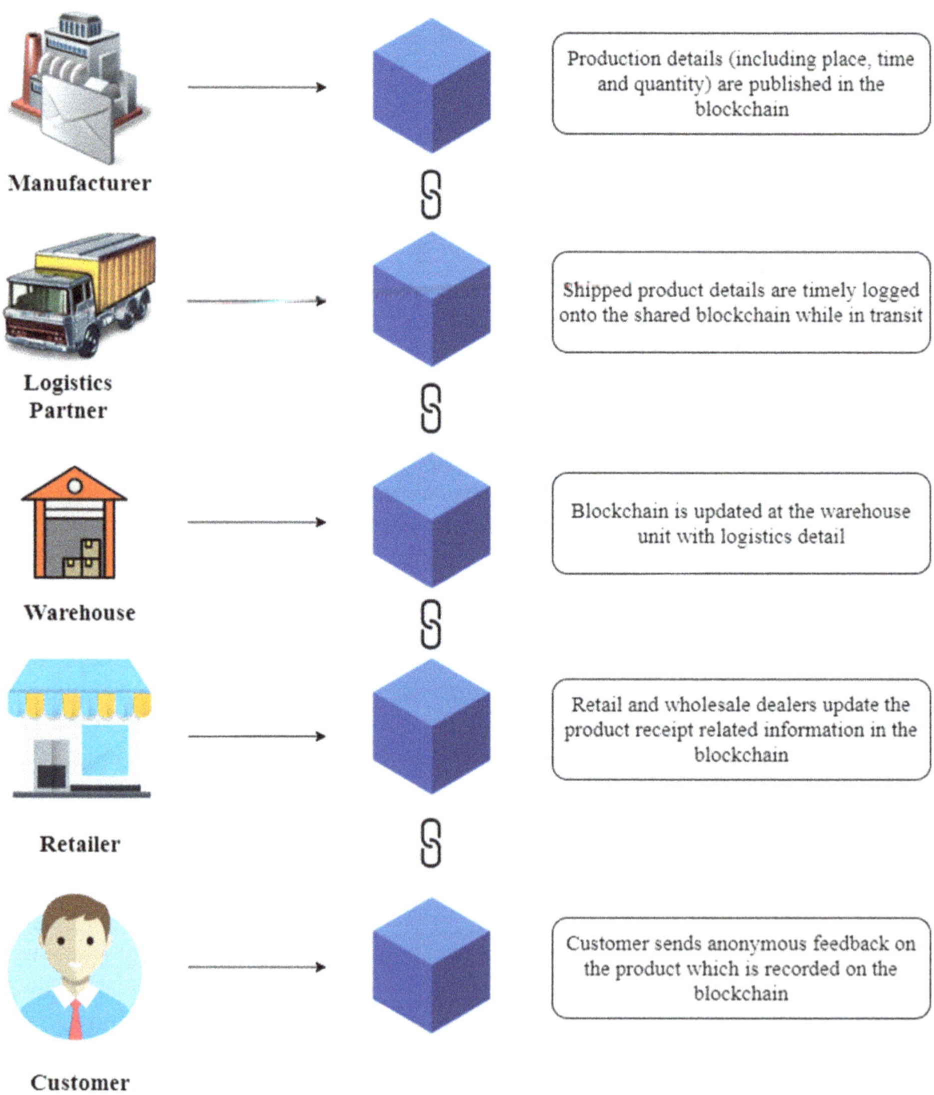

Figure 10.2. Blockchain-based product tracking in the retail industry.

10.8 Summary

Currently, a novel rapid growing trend is the combination of blockchain with IIoT, since a blockchain technique can assist in addressing the challenges of IIoT. Blockchain technology has been widely utilized in many industries. This chapter discussed the applications of BIIoT such as smart manufacturing, management of supply chain, the food industry, smart grid, health, UAV and IoV, E-commerce and the retail industry. Combining blockchain with IoT can bring several advantages to the above mentioned applications. In summary, BIIoT has advantages such as decreasing costs for trusted third parties, guaranteeing safety, enhancing information traceability, checking genuineness of information and protecting confidentiality.

References

[1] Christidis K and Devetsikiotis M 2016 Blockchain and smart contracts for the internet of things *IEEE Access* **4** 2292–303

[2] Kusiak A 2018 Smart manufacturing *Int. J. Prod. Res.* **56** 508–17

[3] Wan J, Li J, Imran M, Li D and e-Amin F 2019 A blockchain-based solution for enhancing security and privacy in smart factory *IEEE Trans. Ind. Inf.* **15** 3652–60

[4] Huang J, Kong L, Dai H N, Ding W, Cheng L, Chen G, Jin X and Zeng P 2020 Blockchain-based mobile crowdsensing in industrial systems *IEEE Trans. Indus. Inform.* **16** 6553–63

[5] Konstantinidis I, Siaminos G, Timplalexis C, Zervas P, Peristeras V and Decker S 2018 Blockchain for business applications: a systematic literature review *Business Information Systems* ed W Abramowicz and A Paschke (Cham: Springer International Publishing) pp 384–99

[6] Kim H M and Laskowski M 2018 Toward an ontology-driven blockchain design for supply-chain provenance *Intell. Syst. Account. Finance Manag.* **25** 18–27

[7] Tapscott A and Tapscott D 2017 How blockchain is changing finance *Harv. Bus. Rev.* https://hbr.org/2017/03/how-blockchain-is-changing-finance

[8] Tian F 2016 An agri-food supply chain traceability system for China based on RFID amp; amp; blockchain technology *13th Int. Conf. on Service Systems and Service Management (ICSSSM)* 1–6

[9] Sander F, Semeijn J and Mahr D 2018 The acceptance of blockchain technology in meat traceability and transparency *Br. Food J.* **120** 2066–79

[10] Bettín-Díaz R, Rojas A E and Mejía-Moncayo C 2018 Methodological approach to the definition of a blockchain system for the food industry supply chain traceability *Computational Science and Its Applications—ICCSA* (Cham: Springer International Publishing) pp 19–33

[11] Lin Q, Wang H, Pei X and Wang J 2019 Food safety traceability system based on blockchain and epics *IEEE Access* **7** 20698–707

[12] Zhang C, Wu J, Zhou Y, Cheng M and Long C 2018 Peer-to-peer energy trading in a microgrid *Appl. Energy* **220** 1–12

[13] Li Z, Kang J, Yu R, Ye D, Deng Q and Zhang Y 2018 Consortium blockchain for secure energy trading in industrial Internet of Things *IEEE Trans. Ind. Inf.* **14** 3690–700

[14] Aitzhan N Z and Svetinovic D 2018 Security and privacy in decentralized energy trading through multi-signatures, blockchain and anonymous messaging streams *IEEE Trans. Dependable Secure Comput.* **15** 840–52

[15] Pop C, Cioara T, Antal M, Anghel I, Salomie I and Bertoncini M 2018 Blockchain-based decentralized management of demand response programs in smart energy grids *Sensors* **2018** 162

[16] Wang K, Shao Y, Shu L, Zhu C and Zhang Y 2016 Mobile big data fault-tolerant processing for e-health networks *IEEE Network* **30** 36–42

[17] Esposito C, Santis A D, Tortora G, Chang H and Choo K R 2018 Blockchain: a panacea for healthcare cloud-based data security and privacy? *IEEE Cloud Comput.* **5** 31–7

[18] Griggs K N, Osipova O, Kohlios C P, Baccarini A N, Howson E A and Hayajneh T 2018 Healthcare blockchain system using smart contracts for secure automated remote patient monitoring *J. Med. Syst.* **42** 130

[19] Bhuiyan M Z A, Zaman A, Wang T, Wang G, Tao H and Hassan M M 2018 Blockchain and big data to transform the healthcare *Proc. of the Int. Conf. on Data Processing and Applications, ser. ICDPA* (New York: ACM) pp 62–8

[20] Sun Y, Zhang R, Wang X, Gao K and Liu L 2018 A decentralizing attribute-based signature for healthcare blockchain *2018 27th Int. Conf. on Computer Communication Networks (ICCCN)* pp 1–9

[21] Rahman M A, Hossain M S, Loukas G, Hassanain E, Rahman S S, Alhamid M F and Guizani M 2018 Blockchain-based mobile edge computing framework for secure therapy applications *IEEE Access* **6** 72469–78

[22] Wu F *et al* 2020 A new coronavirus associated with human respiratory disease in China *Nature* **579** 265–9

[23] Esposito C, De Santis A, Tortora G, Chang H and Choo K-K R 2018 Blockchain: a panacea for healthcare cloud-based data security and privacy? *IEEE Cloud Comput.* **5** 31–7

[24] Wu H-T and Tsai C-W 2018 Toward blockchains for healthcare systems: applying the bilinear pairing technology to ensure privacy protection and accuracy in data sharing *IEEE Consum. Electron. Mag.* **7** 65–71

[25] Salahuddin M A, Al-Fuqaha A, Guizani M, Shuaib K and Sallabi F 2017 Softwarization of Internet of Things infrastructure for secure and smart healthcare *Computer* **50** 74–9

[26] Nugent T *et al* 2016 Improving data transparency in clinical trials using blockchain smart contracts *F1000 Res.* **5** 2541

[27] Xia Q I, Sifah E B, Asamoah K O, Gao J, Du X and Guizani M 2017 MeDShare: trust-less medical data sharing among cloud service providers via blockchain *IEEE Access* **5** 14757–67

[28] Zhang J, Xue N and Huang X 2016 A secure system for pervasive social network-based healthcare *IEEE Access* **4** 9239–50

[29] Vruddhula S 2018 Application of on-dose identification and blockchain to prevent drug counterfeiting *Pathogens Global Health* **112** 161

[30] Alangot B and Achuthan K 2017 Trace and track: enhanced pharma supply chain infrastructure to prevent fraud *Proc. Int. Conf. Ubiquitous Commun. Netw. Comput.* (Bangalore; Cham: Springer) 189–95

[31] Bocek T, Rodrigues B B, Strasser T and Stiller B 2017 Blockchains everywhere—a use-case of blockchains in the pharma supply-chain *Proc. IFIP/IEEE Symp. Integr. Netw. Service Manage (IM)* 772–7

[32] Nchinda N, Cameron A, Retzepi K and Lippman A 2019 MedRec: a network for personal information distribution *2019 Int. Conf. on Computing, Networking and Communications (ICNC)* (IEEE) 637–41

[33] Franks P C 2020 Implications of blockchain distributed ledger technology for records management and information governance programs *Records Manage. J.* **30** 287–99

[34] Yang Z, Yang K, Lei L, Zheng K and Leung V C M 2018 Blockchain-based decentralized trust management in vehicular networks *IEEE Internet Things J.* **6** 1495–505

[35] Kang J, Yu R, Huang X, Maharjan S, Zhang Y and Hossain E 2017 Enabling localized peer-to-peer electricity trading among plug-in hybrid electric vehicles using consortium blockchain *IEEE Trans. Ind. Inf.* **13** 3154–64

[36] Kang J, Yu R, Huang X, Wu M, Maharjan S, Xie S and Zhang Y 2019 Blockchain for secure and efficient data sharing in vehicular edge computing and networks *IEEE Internet Things J.* **6** 4660–70

[37] Liu H, Zhang Y and Yang T 2018 Blockchain-enabled security in electric vehicles cloud and edge computing *IEEE Netw.* **32** 78–83

[38] Dai Y, Xu D, Maharjan S, Qiao G and Zhang Y 2019 Artificial intelligence empowered edge computing and caching for the internet of vehicles *IEEE Wirel. Commun. Mag.* **26** 12–8

[39] Zeng Y, Zhang R and Lim T J 2016 Wireless communications with unmanned aerial vehicles: opportunities and challenges *IEEE Commun. Mag.* **54** 36–42

[40] Kimchi G, Buchmueller D, Green S A, Beckman B C, Isaacs S, Navot A, Hensel F, Bar-Zeev A and Rault S S J-M 2017 Unmanned aerial vehicle delivery system *US Patent* 9,573,684

[41] Wang L, Chen F and Yin H 2016 Detecting and tracking vehicles in traffic by unmanned aerial vehicles *Autom. Constr.* **72** 294–308

[42] Cheng N, Xu W, Shi W, Zhou Y, Lu N, Zhou H and Shen X 2018 Airground integrated mobile edge networks: architecture, challenges, and opportunities *IEEE Commun. Mag.* **56** 26–32

[43] Li B, Fei Z and Zhang Y 2019 UAV communications for 5G and beyond: Recent advances and future trends *IEEE Internet Things J.* **6** 2241–63

[44] Kapitonov A, Lonshakov S, Krupenkin A and Berman I 2017 Blockchain-based protocol of autonomous business activity for multiagent systems consisting of UAVs *2017 Workshop on Research, Education and Development of Unmanned Aerial Systems (RED-UAS)* 84–9

[45] Kumar A, Kundu A, Pickover C A and Weldemariam K 2018 Unmanned aerial vehicle data management *US Patent App.* 15/463,147

www.ingramcontent.com/pod-product-compliance
Ingram Content Group UK Ltd.
Pitfield, Milton Keynes, MK11 3LW, UK
UKHW051850100926
8556IPUK00005B/549

9 780750 336611